APPLIED MATHEMATICS

Applying Number and Quantity to Everyday Life

Erik Richardson

New York

Published in 2017 by Cavendish Square Publishing, LLC
243 5th Avenue, Suite 136, New York, NY 10016

Copyright © 2017 by Cavendish Square Publishing, LLC

First Edition

No part of this publication may be reproduced, stored in a retrieval system, or transmitted in any form or by any means—electronic, mechanical, photocopying, recording, or otherwise—without the prior permission of the copyright owner. Request for permission should be addressed to Permissions, Cavendish Square Publishing, 243 5th Avenue, Suite 136, New York, NY 10016. Tel (877) 980-4450; fax (877) 980-4454.
Website: cavendishsq.com

This publication represents the opinions and views of the author based on his or her personal experience, knowledge, and research. The information in this book serves as a general guide only. The author and publisher have used their best efforts in preparing this book and disclaim liability rising directly or indirectly from the use and application of this book.

CPSIA Compliance Information: Batch #CS16CSQ

All websites were available and accurate when this book was sent to press.

Library of Congress Cataloging-in-Publication Data

Names: Richardson, Erik.
Title: Applying number and quantity to everyday life / Erik Richardson.
Description: New York : Cavendish Square Publishing, [2017] | Series: Applied mathematics | Includes bibliographical references and index.
Identifiers: LCCN 2016000266 (print) | LCCN 2016010752 (ebook) | ISBN 9781502619631 (library bound) | ISBN 9781502619648 (ebook)
Subjects: LCSH: Mathematics--Popular works.
Classification: LCC QA93 .R53 2017 (print) | LCC QA93 (ebook) | DDC 510--dc23
LC record available at http://lccn.loc.gov/2016000266

Editorial Director: David McNamara
Editor: B.J. Best
Copy Editor: Nathan Heidelberger and Rebecca Rohan
Art Director: Jeffrey Talbot
Senior Designer: Amy Greenan
Production Assistant: Karol Szymczuk
Photo Researcher: J8 Media

The photographs in this book are used by permission and through the courtesy of: James Galpin/Moment/Getty Images, cover; STILLFX/Shutterstock.com, 4; Milan Gonda/Shutterstock.com, 8-9; Leemage/UIG/Getty Images, 14; LAGUNA DESIGN/Getty Images, 18; Radu Razvan/Shutterstock.com, 20; Aluxum/Vetta/Getty Images, 24; Kevin Elvis King/Moment Select/Getty Images, 32; Satyrenko/Shutterstock.com, 35; Sorincolac/iStock/Thinkstock, 38; AaronChenPs/Moment/Getty Images, 40; ra-photos/E+/Getty Images, 42-43; Alfred Edward Chalon/File:Ada Lovelace portrait.jpg/Wikimedia Commons, 46; Tom Yates/File:Bombe-rebuild.jpg/Wikimedia Commons, 50; Science Museum London, 53; Niklaus Bernoulli/File:Jakob Bernoulli.jpg/Wikimedia Commons, 54; File:KochFlake.svg/Wikimedia Commons, 69; Pavel Ignatov/Shutterstock.com, 73; Ferdinand Schmutzer, restored by Adam Cuerden/File:Einstein 1921 by F Schmutzer - restoration.jpg/Wikimedia Commons, 74; Nick Risinger/File:Milky Way Galaxy.jpg/Wikimedia Commons, 78-79; Majeczka/Shutterstock.com, 81; NASA/JHU APL/SwRI/Steve Gribben/File:15-011a-NewHorizons-PlutoFlyby-ArtistConcept-14July2015-20150115.jpg/Wikimedia Commons, 86; Photographee.eu/Shutterstock.com, 88; John Lund/Blend Images/Getty Images, 95; Georgios Kollidas/Shutterstock.com, 96; Ambrozinio/Shutterstock.com, 102-103; Zhu Difeng/Shutterstock.com, 110-111; Encyclopedia Britannica/UIG Via Getty Images, 117.

Printed in the United States of America

TABLE OF CONTENTS

5   Introduction

7   ONE **A History of Number and Quantity**

21   TWO **Number and Quantity in Your Everyday Life**

81   THREE **Number and Quantity in Others' Everyday Lives**

115   Conclusion

118   Glossary

121   Further Reading

124   Bibliography

126   Index

128   About the Author

Money itself is a kind of model—a way we translate objects and activities around us into numbers so we can make comparisons and decisions.

INTRODUCTION

When do you think you first thought about math and counting? Were you four? Maybe two? What if I told you it was even earlier?

There have been studies that show that even newborns seem to have some awareness of the differences between one thing, two things, and three things. This makes it seem like math and numbers might be built into our brains, but it's not just humans.

Various kinds of animals seem to be able to count at a basic level. This includes a wide range, from magpies and pigeons to rats and monkeys. There have even been ways that scientists figured out to test whether ants use counting to help them remember where their anthill is. If the scientists lengthen or shorten sticks that have been attached to the ants' legs—thereby making their steps travel farther or shorter distances with the same number of steps—the ants overshoot or undershoot the distance back to the anthill.

As I heard it explained once, if you don't think your dog can count, show them you are putting two treats behind your back, then give them only one. See how sure they are that there's one still pending.

This rudimentary counting is fascinating and would suggest that math is woven into the very fabric of mental processing, even at the basic level of ants. But the point at which it really becomes significant is where the process begins to move beyond mere counting. It is the point where we instead start looking for new ways to shorten up the counting or other processes that we are trying to accomplish, or where we try to stretch the shortcuts and patterns we've already identified to get them to apply to a new, more complex problem or situation.

    Heading into this book, that is the one key idea that you should take with you: the goal of math is to make our lives easier! I know it can feel far from that sometimes—especially when confronted with a long or challenging math problem or assignment. Yes, it might be a long, hard math problem, I'll admit, but it is always shorter or easier or less dangerous than the way we would accomplish that thing by going the long way. Working with decimals may seem hard, but consider how long it would take to figure out how much money the bank owes you if you had to actually stand there and move the pennies around between one pile and the next every time you put money in or took money out. Want to design and play a cool video game? If you want it to be realistic *and* be ready to test out before you are old and gray headed, you're going to need a ton of shortcuts, and you'll see that math can provide them.

ONE

A History of Number and Quantity

This is a chapter about the history of number and quantity and mathematics that is not going to actually be about the whole picture. It is quite common to pick up a general discussion of math and see the authors launch into a discussion of which primitive tribes have the oldest archaeological evidence that they counted or marked sticks to measure stuff, or a bunch of other similar things that you and I probably don't care about all that much. Starting a discussion of the history of math by talking about the birth and spread of counting is kind of like starting a book on the history of archery by spending the first chapter talking about which culture picked up sticks first.

If math is anything, it is a collection of tools and strategies for allowing us to find a shorter and more interesting way to get where we are going without having to slog through long, slow parts. Why not start right here in chapter one by skipping around the long, slow march over thousands and thousands of years of basic counting with rocks and marked sticks and beads?

Greece was the birthplace of much of our thinking about math. We can see math expressed through ancient Greek architecture, such as the Acropolis.

8   Applying Number and Quantity to Everyday Life

A History of Number and Quantity 9

Instead, let's talk about some cool, exciting moments where math suddenly leapt forward—some moments where our minds stretched out and tried to represent some things that we had not yet seen, or tried to think in a way we had not tried to think before. Those are the cool things. They are the answer to questions like "Why do we have to learn this stuff?" and "What is this good for?"

I'm not going to go through all the little steps in between where parts of ideas developed but didn't catch on, or where somebody had a sort of good idea that gave someone else an even better idea. It's true: math, like science, happens in thousands of little baby steps, with an occasional big step that seems like a giant leap. However, one of the tricks we've learned from math itself along the way is that trying to figure out all the tiny steps in between can make your brain hurt and can keep you from seeing how the big pieces actually fit together.

So … let's meet some new ideas and pretend like we are jumping from boulder to boulder to travel across the river. If you feel an overwhelming urge to find out a little more about how a particular breakthrough actually unfolded, you should check out a book, or look it up on a website—such as those given at the back of this book!

1. Euclid owned geometry (like a boss).

While early civilizations like the Egyptians and Babylonians used some **geometry** for practical problems like measuring their fields, it was the Greeks, starting around 575 BCE, who really tried to figure out a way to put down a set of rules that would allow them to solve groups of problems and to prove that the answers were dependable. They wanted to do this without physically measuring every small square of the field or every round pillar holding up a temple to check their answers. This process reached a kind of peak with the great mathematician Euclid around 300 BCE. The textbook he wrote, *Elements*, was the standard textbook for geometry for the

The Tortoise and Achilles

Unfortunately, this story does not seem to have survived in Zeno's original writings, but it was recounted in some of the works of Aristotle. The basic idea is this:

If the tortoise is given a head start, then even the fastest runner from mythology would not be able to catch him. In the few seconds it will take Achilles to catch up to the tortoise, the tortoise will have moved a small distance forward while Achilles was catching up to him. Maybe the tortoise is 1/20th as fast as Achilles. So, while Achilles ran the distance of 10 meters, the tortoise only ran an additional 0.5 meters. So now Achilles is only 0.5 meters behind, and he runs again, but the tortoise also runs a few slow centimeters forward. Then Achilles runs that 0.5 meters, but the tortoise moves forward just a bit more. It seems every time Achilles tries to catch up, the tortoise will always be just a bit ahead.

The interesting thing about this (and some of Zeno's other ideas as well) is that while it seems like some sort of riddle, it is, in fact, evidence of an amazing intellect. The **paradox** this story created has troubled and tinkered around in minds of mathematicians, logicians, and philosophers all the way up to the present. It is easy to claim we can refute it, as slow runners are caught and passed by faster runners all the time. But can you construct a rigorous **proof** that your methods work?

A History of Number and Quantity

next two thousand years, and it holds the all-time record for most successful math book.

2. Zeno came up with some paradoxes that hurt everyone's brains.

Also during the period of Greek advances in mathematics, a clever thinker named Zeno, around 450 BCE, was able to clearly set out a group of puzzles, or paradoxes, that helped to show that trying to apply math and measurements to the normal world around us sometimes doesn't work. If you read the sidebar about why a tortoise would be able to beat Achilles, a hero from Greek myths, in a race, you will appreciate how good of a puzzle it is. You might think to yourself that an actual race could prove the answer, but the challenge is to figure out how you would prove it just using math. If we have to go out and do everything by hand to find the results and make sure they're dependable, we haven't saved ourselves any time or work.

3. Hindu-Arabic numbers spread to the West.

Born around 1170 CE in Pisa, Italy, Fibonacci had the great fortune to spend much time in the area that is now Algeria. Traveling there with his father, a merchant, Fibonacci was able to study with scholars and learn the Hindu-Arabic system of numbers and operations, which were a significant improvement over the cumbersome Roman numerals still in use in the late 1100s and early 1200s.

It was Fibonacci's book, *Liber Abaci*, published in 1202, that went on to introduce the Hindu-Arabic numbering system and operations to Europe, including the use of place value and the closely related importance of zero in arithmetic.

4. We figured out how to do math with important pieces of the question missing.

From early on, our math was about solving problems where the answer was missing, but we didn't have a systematic, careful way to solve problems when we knew something about the answer but part of the question was missing. Like geometry and other areas of math, the gradual stages when different pieces fell into place are messy and complicated. We can point to a key moment in the ninth century CE, though, when a mathematician from Persia (modern Iran), named Muhammad ibn Musa al-Khwarizmi, wrote the text that translates as *The Compendious Book on Calculation by Completing and Balancing* and formulated the process into a system parallel to the one used by Euclid. He reduced parts of equations to standardized forms, such as:

$$ax^2 + bx = c, (a + b)^n$$

The word "algebra" is from the Arabic, and means "restoration." However, it was not until the sixteenth century that the ideas in the text created an explosion of interest in algebra. It took off in the West and important advances were made which increased its usefulness to a wider range of problems.

5. Descartes came up with a plan so algebra and geometry could get married.

Rene Descartes, a French mathematician born in 1596, was an incredibly brilliant guy who single-handedly created a revolution in the field of philosophy and another one in the field of mathematics. His great breakthrough in mathematics was to think

Rene Descartes's ideas were like a kind of telescope or microscope—he turned piles of numbers into pictures so we could see them.

Applying Number and Quantity to Everyday Life

Descartes and the reboot of philosophy

Descartes was equally profound for taking his same manner of careful, systematic thought and applying it to questions beyond the borders of math. By trying to break his beliefs down to the simplest, clearest **axioms**, like those behind the work of Euclid and al-Khwarizmi, he hoped to clear away much of the clutter and confusion that had gradually accumulated in the Middle Ages with respect to questions about why we are here, what our purpose is, could we know if there is a God, and so on.

One of his most brilliant ideas was to realize that almost any of his ideas or things he thought were true might be a mistake, like the way someone who has made an error in their math homework "thinks" they are right.

The insight he came up with was that even if he was wrong, even if he were suffering from illusions, it would still be true that *he* was real—otherwise, how could he be deceived? That is what gave rise to his most famous claim, "I think, therefore I exist."

His ideas were so profound, and his mathematical style of thinking so inspiring, that other great minds of Europe adopted his approach, and while there have been developments and splits in the time since Descartes, his work is still studied in colleges around the world.

of a way that we could take algebra problems and map them out on a **plane** so that we could see and interpret different things about them. That way of mapping algebra equations is still called the Cartesian coordinate system. Before Descartes, no one had ever seen anything like it.

Fermat, another brilliant mathematician, came up with a similar idea at the same time, though their developments were unknown to each other. This kind of thing happened from time to time in the olden days, when communicating was slow, and it sometimes made it hard to know how to give proper credit to this inventor or that one. A similar thing happened in the case of calculus.

6. What do we do if something isn't true, but only "probably true"?

In the case of **probability** and statistics, the key turning point, when work from previous thinkers came together in a clear, organized way, was the work by Jakob Bernoulli, though the text he wrote wasn't published until 1713, after his death. We might expect, for instance, that if we flip a coin a bunch of times, it will come up heads about the same amount of times as it comes up tails. Figuring out a way to *prove* that is another trick, though!

It is worth noting that two of the mathematicians doing serious work in probability leading up to Bernoulli's proof were Blaise Pascal and Pierre de Fermat. They are both giants of the math realm, and we will meet them later in this book and in other books in the series.

7. Newton and Leibniz captured infinity and allowed us to finally beat Zeno's puzzles.

While math had made great progress in many areas since the time of the ancient Greeks, it had made very little meaningful progress on the puzzles Zeno had set out. While math was good at finding areas of regular shapes and figuring out certain things involving problems that can be represented as a straight line, it had not yet come up with a helpful way to deal with shapes and graphs that had **irregular** curves or with problems involving an infinite series of things or values.

Then, in the late seventeenth century, at the same time but unknown to each other, two brilliant characters invented calculus. It's hard to describe how it works in a short amount of space, but think of it as a way to get an approximate area by slicing the shape (or the graph) into pieces that are almost zero inches (or millimeters) wide, and as a result, coming up with an approximate answer that has almost zero amount of error in it.

8. Gödel proved that we can't prove all math stuff is true, even if it *is* true.

In the late 1800s and early 1900s, there was a big effort to try to get all of math to fit into a rigorous logical framework, like had happened with the smaller area of geometry in the work of Euclid in ancient Greece. Incredibly smart people like David Hilbert, Bertrand Russell, Alfred North Whitehead, and Gottlob Frege (one of the smartest people in logic since the time of Aristotle!) worked on these problems. It was a hard project, and several of them got really close, all the while debating back and forth with each other to find and fix holes in their different attempts.

Then along came the German mathematician Kurt Gödel, who created a kind of proof so awesome that no one had ever thought of it before and it really blew the other people out of the water. Once they understood his proof, they all saw that any system with enough parts to do basic mathematics would always be capable of generating principles and formulas which could be true, but which that system could not prove were true. To give you just a little feeling for it, it shows that a system would always be able to create a **proposition** that says, "This statement cannot be proved." If the system could prove it, then it is false, so the system generated a

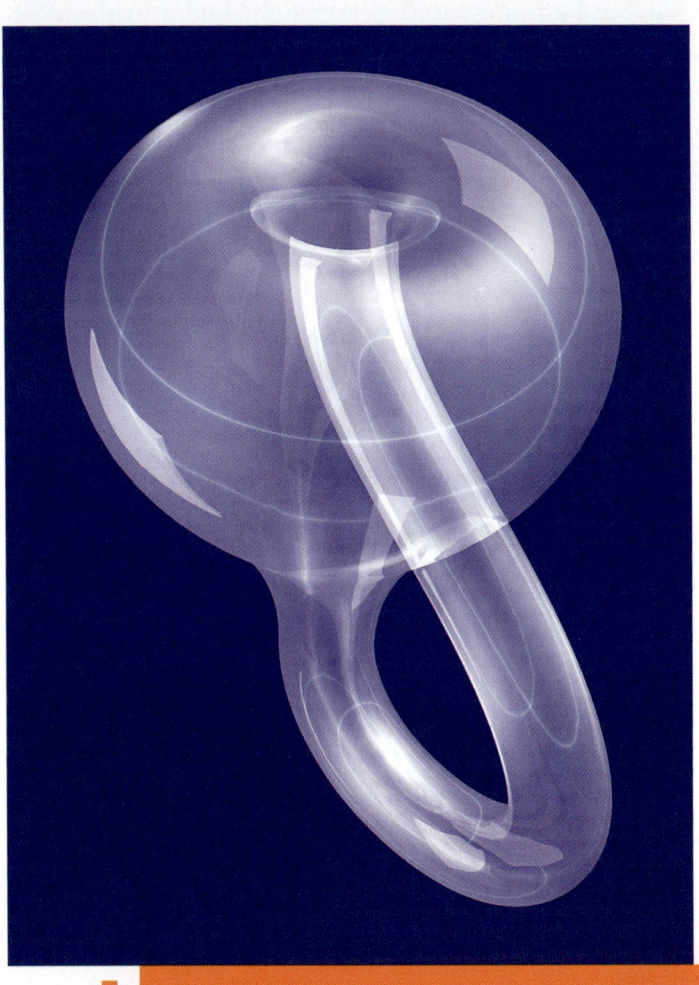

This is an example of a Klein bottle. It's a special kind of mathematical object where the inside and outside are all one surface.

false statement and can't be trusted. If the system cannot prove it, then it is true, but the system is not capable of proving it. To be fair, you might have seen this kind of paradox before, but the brilliance was to show that any set of math rules and equations would create a statement that has the same effect.

9. Someone dropped regular geometry and broke it, and new kinds fell out.

In contrast to the way calculus solved a set of problems that Zeno had created, in this case some modern mathematicians took a solution Euclid had come up with and unsolved it. Part of the goal and the great contribution of Euclid's *Elements* was to create a method of proving different things about geometry by showing they were built from a few basic rules. One of these rules that seemed very obvious to everyone for the last two thousand years had always been kind of a troublemaker. It was hard to show how it could be built using just the other rules and definitions. So, a couple clever math guys, again not knowing the other was doing it, decided to test out what would happen if they got rid of the one rule that was causing trouble. The result was that without that rule tying them down, they could use the others to build a set of proofs and procedures that would allow for understanding how geometry works on surfaces that aren't flat—like globes, for instance (which comes in pretty handy if you happen to bump into anyone who lives on a giant globe).

These new kinds of geometry—called, helpfully enough, non-Euclidean geometries—would turn out to be very important for understanding the bent, warped, and dented shape of space-time that was central to Einstein's work later on.

So, with this quick tour through some of the high points in the history of math, we have a kind of general overview. Like math models themselves, it is simpler than the real thing, but that has to happen in order for it to be useful. Let's turn, then, to start exploring why it's a big deal whether we figured out how to do this or that. The short answer has two parts. Part one is that math is kind of like the blueprint for figuring out the world around us, from teeny-tiny stuff like bacteria to enormous stuff like galaxies. Part two is that math turns up in a ridiculous number of different jobs, sports, and games all over the place as people use it to help them get better at the activities they are doing.

Cycling is an example of a sport where there are numbers working just below the surface. For example, we have the revolutions of the wheels, the incline of the hill, the gear ratios, and more.

TWO

Number and Quantity in Your Everyday Life

In this chapter, we will spend some time looking at the different places and different ways that number and quantity form a kind of framework for our lives. Hopefully you will be able to look at it with fresh eyes knowing there is not a test waiting at the end of the book, and knowing that you picked this up because you have some curiosity rattling around that wants to crash into some new ideas as if you were riding in the bumper cars at an amusement park. I think the wide assortment of topics will help you to find at least a few things that spark you to go on and be curious on other days and on other math topics, too.

Sports and Recreation

While we often notice numbers at the center of different sports, in things like the scores, race times, rushing yards, batting averages, and pass completion percentages, numbers and quantity are also deeply built into sports in ways that we *don't* usually notice. Numbers and quantity are woven into the fabric of many different sports.

Consider, for instance, the math at play when we hear that familiar crack of the bat if the pitcher puts it over the plate a little too slow or a little too straight. The same math applies to a knockout punch or kick in the mixed martial arts octagon.

When an object is moving, it has **kinetic** energy. When it smacks into something, that energy is transferred into whatever it hits. This kinetic energy is measured in **joules**. All we really need to understand here, though, is that in everyday language 1,000 joules would be about as much energy as dropping a 22-pound (10-kilogram) block out of a third-floor window.

When we want to calculate the amount of force delivered by a moving object, we apply the following equation:

$$\text{Kinetic energy} = \tfrac{1}{2} \times \text{Mass} \times \text{Velocity}^2$$

In that equation, "**mass**" is basically "how much matter there is in something." We find how much by taking the weight of the object (in pounds) and dividing by the pull of gravity (32.2 feet per second per second). "**Velocity**" is the change in position (how far it travels) divided by the time over which the change occurs (the rate of speed).

Based on different studies, we know that a good boxer or martial artist throws a punch with mass between 6 and 9 pounds (between 2.7 and 4.1 kilograms), and at impact it is traveling at between 9 and 11 meters (30 to 36 feet) per second.

Running our calculation, then, we'll pick some values in between those ranges, and we would see that the kinetic energy of the punch is:

$$KE = \tfrac{1}{2} \times 4 \text{ kg} \times (10 \text{ m/sec})^2,$$
which comes out to 200 joules.

By comparison, if we were hit in the head with a baseball pitch traveling at 100 miles per hour (44.7 meters per second), we would only have $\tfrac{1}{2} \times 0.142 \text{ kg} \times (44.7 \text{ m/sec})^2 = 142$ joules.

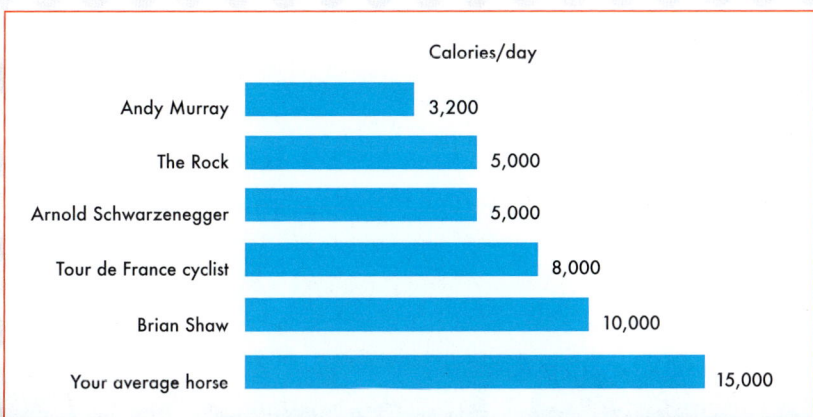

World's strongest men

Understanding eating requirements is crucial to the people who compete in the World's Strongest Man competitions, like Brian Shaw, a four-time winner.

Let's think about the calories they consume in terms of the amount of work they do in a day of training.

Brian Shaw consumes 10,000 calories per day, which is about the energy contained in 0.32 gallons (1.21 liters) of gasoline. If your car gets 30 miles to the gallon (12.75 kilometers per liter), then these strong men are working out about as hard as if they pushed your car along the highway for 9.6 miles (15.4 kilometers) every day!

When you see the results, though, it is truly impressive. With an interesting and entertaining range of events—from the keg giant dumbbell press, to the log lifts, to the events where contestants have to pick up a modified car frame that they are standing inside and carry it down the field—the competition is exhausting to even watch! Perhaps the most unconventional and most intimidating event is the Atlas Stones, where giant stone spheres have to be hoisted up onto massive columns at about mid-chest height.

The events vary to prevent over-specialization. Sometimes there is an event where they have to pull a diesel cab. Seems pretty fitting, given the point about how far they could push your car!

Number and Quantity in Your Everyday Life

A punch from a trained boxer contains much more kinetic energy than a baseball pitched at top speed.

Whether an athlete is working to increase speed or build strength, they must use calculated nutrition programs. Athletes work carefully to scientifically manage their calorie levels and types as they do their training routines and recovery periods.

Let's take bodybuilders nearing a contest as an example. As the contest approaches, they will be working to maintain muscle mass and reduce any excess body fat to help their muscles stand out. An important **baseline** measurement is to determine the number of daily calories needed for their given size and activity levels.

As they adjust that calculation to allow for a slow decrease in body fat, they must also work to make sure a certain amount of those calories come from protein, fat, and carbohydrates. For example, Arnold Growingbigger has a competition weight of 235 pounds and consumes a daily calorie level of 5,000 calories per day as he nears a competition event.

With a target of 40 percent of calories from protein, 40 percent from carbs, and the remaining 20 percent from fats, then Arnold would aim for:

$$5{,}000 \text{ calories/day} \times 0.40 = 2{,}000 \text{ calories per day from protein}$$

$$5{,}000 \text{ calories/day} \times 0.40 = 2{,}000 \text{ calories per day from carbohydrates}$$

$$5{,}000 \text{ calories/day} \times 0.20 = 1{,}000 \text{ calories per day from fats}$$

Since 1 gram of protein has 4 calories, that comes out to 500 grams of protein per day. That is the equivalent of 83 eggs or 4 pounds of steak!

Mountain climbing

As runners and other athletes know, one of the key limiting factors to achieving your highest **potential** has to do with the rate your body takes in and uses oxygen, which it needs as part of the process of converting your body's stored energy and changing it into a form your muscles can use.

VO_2 max is a measurement of how much oxygen your body is able to use per minute when exercising at your peak level. In the full-scale test in a lab, you would be hooked up to a breathing apparatus to measure your oxygen intake and output as you run faster and faster on a treadmill. By running calculations based on your rate of oxygen consumption during the exercise test, your heart rate, and your body weight, a physiologist is able to calculate how much oxygen you were consuming when you were running at your hardest. This number is your VO_2 max. While it's difficult to directly use such equipment due to cost and accessibility, there have been several studies done that can provide usable estimates of this metric for training purposes.

One of those is the Uth-Sorensen-Overgaard-Pedersen estimation. The formula is:

$$VO_2 \text{ max} \approx 15 \frac{mL}{kg \times min} \times \frac{HR_{max}}{HR_{rest}}$$

In this formula:

HR_{max} = Your maximum calculated heart rate.
HR_{rest} = Your resting heart rate.

Another common formula that can be used to provide an estimate was developed by Kenneth Cooper for the US Air Force back in the late 1960s. The Cooper test measures the distance covered running in 12 minutes. Based on that distance, an estimate of VO_2 max in mL/(kg × min) is:

$$VO_2 \text{ max} \approx (35.97 \times dmiles_{12}) - 11.29$$

where:

$dmiles_{12}$ = The distance in miles covered in 12 minutes.

By way of comparison, the average healthy male with little or no training will have a VO$_2$ max around 35–40 mL/(kg × min). The average healthy female with little or no serious training will score a VO$_2$ max around 27–31 mL/(kg × min).

In sports where endurance is an important component in performance, like cycling, swimming, and running, world-class athletes tend to have high VO$_2$ max levels. Male runners at the top level can be up around 85 mL/(kg × min), and female runners at that level can be up around 77 mL/(kg × min). In few activities, though, does this become so critical a factor as in the case of mountain climbing.

As you travel higher and higher above sea level, the air **pressure** decreases. That means it is spread out more and more. Your body doesn't actually care how much air you breathe in; what it cares about is how much oxygen you are pulling in and passing along to the muscles to help them do their job. At higher elevations, you are bringing in less oxygen with your breathing, so while your lungs might expand as much as usual, the maximum amount of oxygen you're able to absorb goes down.

Think of having a handful of pieces of chocolate. If that is how much energy you need to take in to perform some task, like walking around a baseball field, then think about how much harder it would be if the chocolate was spread out around the baselines instead of being in your hand. It would become more challenging to get to the chocolate you need using the energy from the last little piece of chocolate. Now imagine that the same handful is spread out around the whole baseball field—infield and outfield. See the challenge?

For many mountain-climbing expeditions, then, the planning process must carefully calculate the oxygen availability level and the climber's oxygen requirements, and then plan to bring along the right amount of additional oxygen to make those numbers balance.

Of course, how much gear is being carried will change how much energy (and how much oxygen) is required, so the gear weights must be managed—this includes the extra weight of carrying the bottles of oxygen. Imagine if carrying an extra piece of chocolate took as much energy as eating the chocolate would give you. It's an even trade. What if it took just a little more energy? Then what?

Let's consider an example: What if you were at base camp on Mount Everest, in the Himalayas? That camp is at **altitude** 5,364 meters (17,600 feet). A good rule of thumb formula for calculating atmospheric pressure is:

$$p = p_0 \text{ kPa} \times (0.89)^h$$

where:
p_0 = Average pressure at sea level (altitude 0, by definition).
h = The altitude in kilometers (1,000 meters in 1 km).
kPa = Kilopascals, a standard unit of pressure.

If we plug our base camp number into our atmospheric pressure formula, the result would be: $p = 101.3 \text{kPa} \times 0.89^{5.364} = 54.217$ kPa. That means you have about 54 percent as much oxygen as usual per breath (so you're missing 46 percent of the oxygen you normally get per breath—nearly one half!).

Camp 3 is at around 7,000 meters. As you look out from there over the Tibetan plateau, you are working with just 44 percent of your normal amount of oxygen. Then, by the time you get to the summit, at 8,850 meters, you are down to 36 percent of your normal amount of oxygen. Think about hiking up the staircase of an office building and only being able to take a breath on every third step.

At this level, your mental sharpness has fallen to about 30 percent of normal. With even basic math skills slipping away, this is not a good time to try to fix any calculation errors you made when you planned out the climb and the oxygen bottles you would need to help you survive.

To compensate, your breathing rate has to increase a great deal, even at rest. Because your body might *over*compensate by allowing blood vessels to leak in the brain or lungs (hello, altitude sickness), now almost all climbers use supplemental oxygen to help them make it to summit of mountains like Everest.

A full, 3-liter bottle weighs around 2.5 to 2.6 kilograms. Approximately 1,060 grams of that is oxygen. To balance out the weight versus the oxygen needs, you need to calculate how long each bottle will last during the climb.

The flow control measures the rate you're using oxygen in liters per minute, so it will save you some work (math people love a good short-cut) if you calculate in liters per minute, rather than trying to figure it out as grams. A full, 3-liter bottle holds about 720 liters of oxygen (there's a little bit of difference when filled, so it's good to check the compression against a standard chart when you start the calculations).

Total liters of oxygen per bottle	Liters used per minute	Minutes of oxygen per bottle	Hours of oxygen per bottle	Bottles needed for a 5-hour climbing day
720	1	720/1 = 720	720/60 = 12	0.42
720	2	720/2 = 360	360/60 = 6	0.83
720	3	720/3 = 240	240/60 = 4	1.25
720	4	720/4 = 180	180/60 = 3	1.7

During the climb, you can vary the flows between 1, 2, 3, or 4 liters per minute. Going higher than 4 is possible, but you can actually overdo it and end up with different health problems instead. Now comes the work to figure out how long each of your bottles would last, depending on your flow setting.

Climbers must carefully calculate their oxygen needs to reach the summit! On the other hand, climbing up is just not as exciting for some people as falling down instead. Let's think about the kind of math going on when someone decides to jump out of a perfectly good airplane and experience the thrill of hurtling toward the earth and then pulling the chute at just the right time.

Skydiving

Just as the mountain climber had to balance weight versus oxygen needs, the skydiver wants to try to **optimize** the fun of the jump balanced against the appropriate safety precautions.

When the skydiver jumps out of the plane at high altitudes, she has very little **drag**, because she is not moving very fast yet. Drag is friction or resistance—in this case, from air. With each second that passes, the skydiver will speed up. Since drag depends on the thickness of the **fluid** (yes, air is a fluid), the speed of the thing falling, and the shape of the thing, like most skydivers, she will move into a position with her head pointed down and her arms and legs folded straight together so she can go faster. Now as she starts to speed up, at the end of fifteen seconds, she could be traveling at about 100 meters per second (224 miles per hour). If she were to open her parachute at this point, the sudden increase in drag would cause severe injuries when the lines yank against her.

As the diver nears the ground, she will shift position to one that increases the amount of area by spreading her arms and legs and flattening out (now more of a belly-flop position than a head-first dive). This will slow her down to something closer to 50 meters per second (112 mph). Opening the parachute then decreases the drag even more, maybe to something like 15 meters per second (33.5 mph). That's enough to be a good jolt when she lands, but not injurious.

One of the interesting puzzles that comes up, though, is why the skydiver doesn't just keep going faster and faster all the way down.

$$v = \sqrt{\frac{2mg}{\rho AC}}$$

where:

m = The mass of the sky diver.
g = The acceleration due to gravity, about 9.8 meters per second per second.
ρ = The density of the air the diver is falling through. This is changing as the altitude changes.
A = The projected area of the diver **perpendicular** to the ground.
C = The drag **coefficient**. This number depends on the shape of the diver. The more streamlined her shape, the lower this coefficient.

Let's pick a skydiver who is about 75 kilograms (165 pounds) and we'll estimate the surface area to be 1.02 square meters. The air density is 0.89. The last number we need is the drag coefficient for a skydiver, which is 0.581. How fast is she moving?

So:

$$v = \sqrt{\frac{2 \times 75 \times 9.8}{0.89 \times 1.02 \times 0.581}}$$

$$v = \sqrt{\frac{1{,}470}{0.52743}}$$

$$= \sqrt{2{,}787.1}$$

$$= 52.79 \text{ m/s}$$

Our answer here translates to about 118 miles per hour. This is called the terminal velocity, because at this speed the air resistance pushing against her offsets the amount gravity wants to speed

Changing the shape of your body when skydiving is a magic trick that seems to change how fast time is moving around you.

32 Applying Number and Quantity to Everyday Life

Number and Quantity in Your Everyday Life 33

Meet Julie Milke

Many of us have certain pictures in mind when we think about the amazing work that nurses do, but for most of us, working through problems with fractions and **ratios** probably isn't one of those pictures. It's actually pretty common, as we found when we talked with Julie Milke, a registered nurse.

Thanks for offering to talk with us about your job, Julie. Can you start out by just telling us your actual title and how many years you've been in this position and what kind of training you needed?

I have a bachelor's degree in science and an associate's degree in nursing. I've been in this job for eleven years.

What are some different ways and kinds of math that you use in your job?

I use math to calculate drug dosages. Some are weight based, some based on kidney function—sometimes you have to calculate kidney function based on weight. Intravenous drugs, the ones that are usually administered with a hanging bag that sends fluid into a needle taped in place on your arm, need to be calculated based on a flow rate, so milligrams per minute or milliliters per hour, stuff like that.

Sounds like a lot of pressure not to make any mistakes in your math!

Oh, sure. You need to be accurate with your calculations or you can really hurt someone. If you run an IV too fast, depending on what drug it is, you can cause the tissue to die where the medicine is

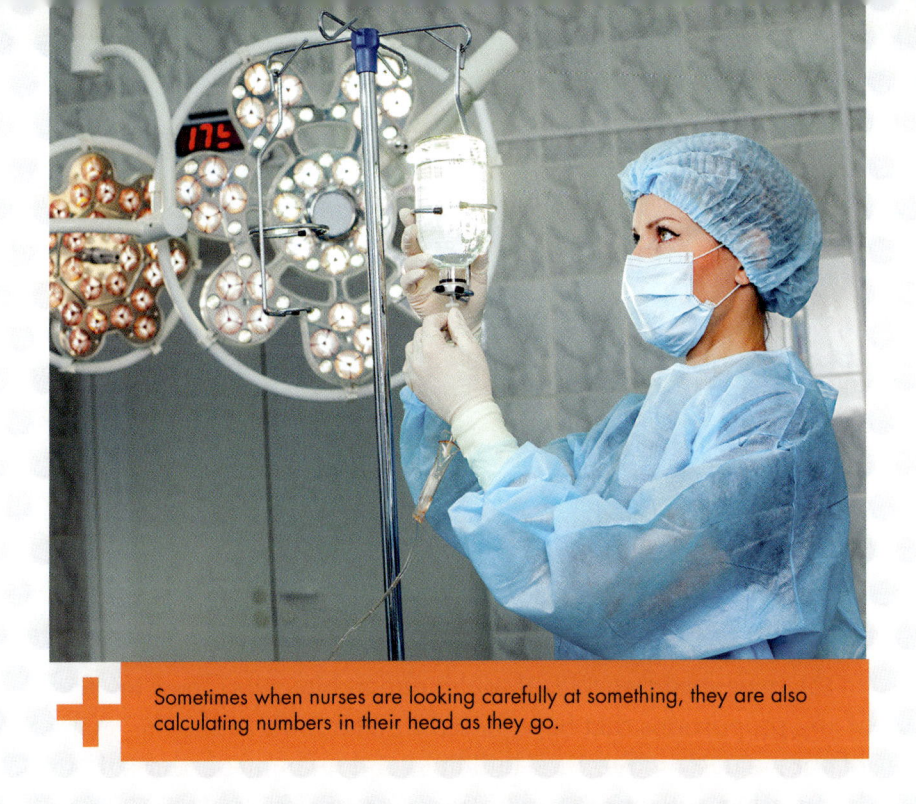

Sometimes when nurses are looking carefully at something, they are also calculating numbers in their head as they go.

going in, or you can easily cause someone's heart to stop, or go way too fast, both of which are, well, obviously pretty bad.

Is there a kind of math problem or application that seems kind of cool compared to "regular old school math?"

Sorry, but I think all the math I have to do *is* regular old school math.

Is there a kind of math or kind of math problem you wish you were better at? Why?

I wish I were better at plain old algebra. It seems like for some people they tend to be better at things like geometry and trigonometry and other people tend to be better at algebra. It's funny, my two kids are complete opposites when it comes to math. One is the "math is easy, I like algebra" type, and the other one is just like me.

her up. Consider what would happen if we changed the area (*A*). Making it smaller would make the overall number bigger—hence allowing her to go faster, something we tend to understand in normal circumstances. The principle is the same for why it feels like you've smacked into something solid if you go for a belly-flop at the swimming pool instead of a nice, sharp dive.

Mountain climbing and skydiving are two activities that require careful preparation to avoid bodily harm. But have you considered the math done by the people who design buildings? Calculations are important to keep those spaces effective, efficient, and safe.

Architecture

One of the oldest ways we see number and quantity come into play in the man-made world is through our architecture. There are few, if any, proportions that have played a greater role in shaping many of our greatest architectural buildings and monuments throughout history than the proportion known as the golden rectangle or the golden ratio.

We see this particular construction reflected in buildings as spread out in time and geography as the pyramids of ancient Egypt, the Parthenon of Greece, Notre Dame Cathedral in France, the Taj Mahal in India, and even the CN Tower in Canada. Across those stretches of time and location, there seems to be a recognition that this proportion "feels" right to us. It doesn't only show up in architecture, but in other forms of art as well. In the realm of painting, we have the *Mona Lisa*, Salvador Dali's *The Sacrament of the Last Supper* (as well as the famous *Last Supper* painting by da Vinci!), and in the work of Mondrian. In music we find not only that intervals and measures of some composers, like Mozart, reflect this proportion, but so does the very design of a violin itself.

What is the golden mean? To understand the proportion we are turning to examine, we must imagine a particular rectangle which has been constructed from a square, as shown above. Starting with

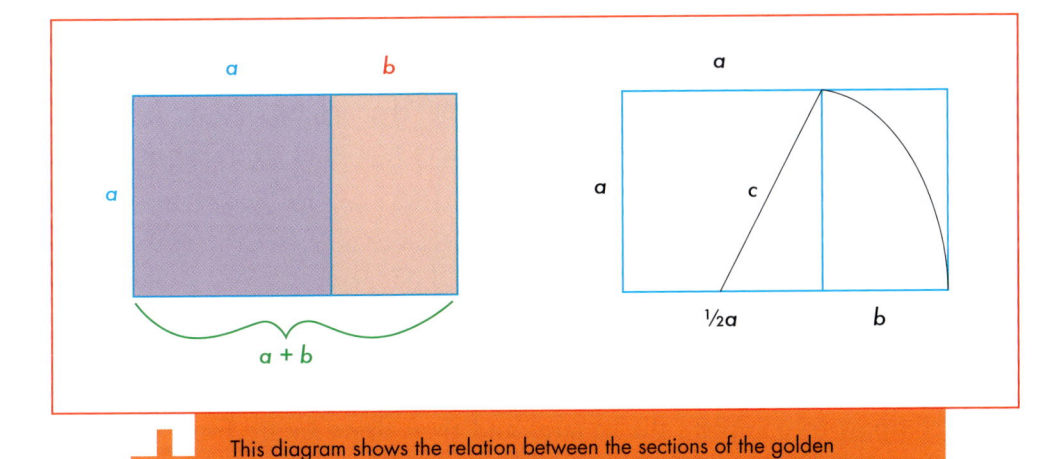

This diagram shows the relation between the sections of the golden rectangle as well as how it is drawn using a compass and straightedge.

a square of $a \times a$ in dimension, you would then find the midpoint of one of the sides.

Place the point of a compass at that midpoint. Choose one of the opposite corners of the square. Draw an arc from that corner outside of the square until the arc reaches the baseline of the square (the horizontal line on which you've placed the point of the compass). You've achieved the golden ratio—the relationship between the dimensions of this particular rectangle. In the diagram, do you think the dimensions look pleasing?

If we let side $a = 1$, then by our handy-dandy Pythagorean theorem, we can find length c.

$$(½a)^2 + a^2 = c^2$$
$$(½ \times 1)^2 + 1^2 = c^2$$
$$¼ + 1 = c^2$$
$$1.25 = c^2$$
$$1.11803 = c$$

If we add that to (½a) we get the length of the bottom of the rectangle, 1.61803.

Then subtracting away *a*, we get the length of *b*. So, 1.61803 − 1 = 0.61803. As a result, there is a kind of mirroring effect, since:

$$a/b = (a + b)/a = 1.618.$$

To construct a decreasing spiral inside and continue the mirroring process to smaller and smaller levels, you can just use a compass to complete the squares and then draw an arc from the far corner, like we did above. With an understanding of what we are seeing, then, let us turn to consider the Parthenon from ancient Greece as one of the **paradigm** examples. By laying an image of the golden rectangle over it, it is easier for us to appreciate the intentions of the Greeks who built it.

This shows how the pattern of the golden mean is captured in the Parthenon, a famous example of Greek architecture.

Rooftop snow loads and surface area

Architects need to be sure to plan buildings whose roofs won't collapse under the weight of snow. In order to find the amount of force being exerted on a roof, we need to start with a calculation for the snow load at ground level. This is usually already done and can be looked up for a given state and region. It is basically a process of taking the amount of snow covering a square foot of ground, melting it down to water, and then weighing it. Typical numbers in northern areas of the United States can range anywhere from 40 pounds per square foot up to 85 pounds per square foot (1,915 to 4,070 pascals).

The roof snow load (S) is given by:

$$S = 0.7GETR$$

where:

G = The ground snow load factor.
E = The exposure factor (like whether there are trees covering/blocking part of the roof).
T = The thermal factor (whether the building is normal; cold/empty, like a warehouse; or really warm, like a factory or greenhouse).
R = The roof slope factor (warm or cold material and how slippery the surface is).

To find the roof slope factor, the equation is:

$$R = (70 - x)/(70 - y)$$

where:

x = The angle of the roof in degrees.
y = The roof type.

Number and Quantity in Your Everyday Life

Brunelleschi not only devised the great dome of the Florence Cathedral but also revolutionized art through the development of linear perspective.

Applying Number and Quantity to Everyday Life

Brunelleschi's dome

Born in Florence in 1377, Filippo Brunelleschi would become famous for solving an architectural puzzle that no one else had been able to crack.

The Medici family was offering a great opportunity to anyone who could finish the huge dome that had been part of the design for the Florence cathedral. Brunelleschi was confident that he could figure out a solution even though he had not been formally trained.

To solve the puzzle, Brunelleschi came up with a new approach that would help distribute the massive weight that such a large dome made of stone would create. He used an inner hemispherical dome, and then a more egg-shaped dome was placed over it.

Brunelleschi had other interesting ideas. In 1434, while the dome was being constructed, he held a public display, sketching the outline of a local building. With a new technique that used reflective material and pinholes, Brunelleschi was able to draw a smaller version of the building in perfect proportions. This allowed him to draw a three-dimensional object in two dimensions much more realistically than had been done before. He had invented perspective. Once the dome was finished, Cosimo de'Medici invited the pope to come and consecrate the cathedral on Easter Sunday 1436. The dome weighed 37,000 tons (33,566 metric tons) and used more than four million bricks. It was a jewel in the skyline of Florence and a sign of the power of the most prominent family.

The minimal-surface roof of the airport terminal in Denver, Colorado, shows how math allows us to solve problems in new ways.

42   Applying Number and Quantity to Everyday Life

Number and Quantity in Your Everyday Life   43

Let's take two sample roofs with dimensions of 100 feet by 120 feet: the Crusty Baker and, across the street, the Caffeine Carnival Gourmet Coffee Shop. Since for most areas the $E = 1$ and the $T = 1$, we can plug those values in to understand the impact. In addition, we will give both buildings a roof type value of 5. The ground snow load factor is 50. The bakery has a flat roof. If overnight weather brings an average snow accumulation for the area, that would give the bakery a load of:

$$S = 0.7 \times 50 \times 1 \times 1 \times [(70 - 0)/(70 - 5)]$$
$$= 37.7 \text{ lbs/ft}^2$$

When we multiply that value times the area of the roof (12,000 ft²), that means there are 452,400 pounds of force that have to be transferred to the walls and foundation. Meanwhile, the sloped roof on the Caffeine Carnival Gourmet Coffee Shop across the street has a 50 degree angle. Its snow load would only be 10.77 lbs/ft², or 129,240 pounds of weight bearing down on the walls and foundation, 28.6 percent as much as the bakery's roof!

Scaling: volume versus surface area

A lot of buildings use domes as part of their architectural design, and while these certainly add interesting balance and proportion, there are more functional reasons that these were developed. Of course, one of the key features of the dome design has to do with the displacement of the weight of the roof itself, but that alone would not have made domes the best solution because, like arches, there is a problem resolving the tension between downward force and the outward force.

To understand the issue that tilts the scale, though, we need to refresh just a bit of basic geometry.

Let us imagine that we are considering a cube-shaped enclosure for an area of our building. The volume to be enclosed is 125,000 cubic feet. In order to do that, we could lay out a floor plan that was 50 feet by 50 feet with a height of 50 feet. In that case, we would need to use enough material to create a surface area of 2,500 × 5 = 12,500 square feet.

In contrast, let us consider using a hemisphere to enclose the space. We're simplifying some things here, since there would likely be a short supporting wall around the base, but that is relatively negligible compared to the larger issue. For a hemisphere that would enclose a volume of 125,000 cubic feet, we could calculate the dimensions of a sphere that would enclose twice that amount, and then we would only use the top half. The formula for the volume of a sphere is:

$$V = (4/3)\pi r^3$$
$$250{,}000 \text{ ft}^3 = (4/3)\pi r^3$$
$$250{,}000 \text{ ft}^3 / (4/3) = \pi r^3$$
$$187{,}500 \text{ ft}^3 = \pi r^3$$
$$187{,}500 \text{ ft}^3 / \pi = r^3$$
$$59{,}683 \text{ ft}^3 = r^3$$
$$39.0796 \text{ ft} = r$$

Let's round it to 39.08 for calculating the next part. The formula for the surface area of a sphere is:

$$SA = 4\pi r^2$$

So, when we plug in the numbers for this case, we see:

$$SA = 4\pi \times (39.08 \text{ ft})^2$$
$$SA = 4\pi \times 1{,}527.246 \text{ ft}^2$$
$$SA = 19{,}191.928 \text{ ft}^2$$

Remember, we are only using the top half of the sphere in this case, so the total surface area we would need to cover now is only 9,595.96 ft^2. That's only about three-fourths as much material as the cube design we considered above. Because a perfect sphere shape does not withstand the outward and downward pressures, most architectural domes you see are modified shapes with a little more height.

We see this same use of math brought to life in the design of various buildings that take the form of geodesic domes. In their case, some of the puzzle of outward versus downward force is offset by the triangular forms of the design. Not only does this allow for use of a lighter-weight material for constructing the shell, but those two things—the shifting of loads through triangles and the lighter material—allow many of them to retain a more fully round shape.

Another interesting way to solve these same puzzles of enclosing a given building size, displacing loads on the roof, and minimizing the materials needed is accomplished by something called minimal surfaces.

While the math is more complex than we can attempt to walk through here, it boggles the mind a little to realize that the mathematics involved in solving this kind of puzzle is worked out through analyzing the behavior of soap bubbles and the way soap film behaves when it is used to cover a wire frame.

The simple flat circle shape you might remember from dipping a bubble wand into a jar before blowing on it is just the beginning! Just imagine, for instance, what might happen if you dipped a spring into a big jar of soapy water. In that case, the spring might behave like a strange sort of spiral staircase—well, more of a spiral slide, actually, as the soap film tries to find the smallest way to stretch across the strange space between the coils of the spring.

Computers and the Internet

If we want to understand just how thoroughly number and quantity are woven into the fabric of our lives, we need look no further than the computer on our lap or the cell phone in our pocket. Our computers practically breathe numbers in the form of **binary** code.

Binary code is the language consisting purely of long strings of 1s and 0s linked together and used to encode just about any kind of information we can think of. On occasion, computer programs even use the strings of 1s and 0s to imitate our own thinking processes.

So how does a stream of electricity coming into the shiny box turn into a complicated string of 1s and 0s, and why? Let's think about it this way: we have a complex language with words and numbers that we string together in long sequences to create different meanings and messages. As soon as you stop and think about our language that way, suddenly what the computer is doing only sounds *mostly* strange instead of, perhaps, *completely* strange. If you think about the fact that we're really just stringing together letters to make the words, then the only real gap left between the way we create meaning and the way computers do would seem to be the distance between letters and those 1s and 0s.

It will be easier to start with numbers, and then letters. We normally count in groups of ten, and each digit in a number represents how many groups of ten smaller bundles we've included. So, for instance, the number 234 indicates that we have four singles. Every time we get ten singles, we bundle them together and pass them over to the left. It appears we've gotten three bundles of ten singles thus far. Similarly, when we get ten of those bundles, we tie them together and pass them up to the left. We have done this two times so far in the number we're considering.

Now, let's imagine that instead of bundles of ten, we had a system that operated with bundles of two. If we get two singles, we tie them together and pass them up to the left. We could represent that as 10, which shows we now have 0 singles and 1 group of two. That's the number 2 in binary, the language of computers. If we have 101 in binary, this indicates that we have 1 single, 0 groups of two, and 1 group of two bundles of two. We understand the number as $(1 \times 1) + (0 \times 2) + (1 \times 4) = 5$. Can you figure out what 1101 in

It's amazing to realize that a computer like Turing's "bombe," one that changed the course of World War II and all of history, had less power than the one sitting on your desk.

48 Applying Number and Quantity to Everyday Life

Turing and imitation

Alan Turing was a brilliant mathematician who first became famous when he solved a big problem in mathematics at the age of twenty-three. The problem had stumped the great mathematicians of the time, and in order to solve it, he came up with the idea of a kind of universal machine that could imitate, by using symbols, any operation that other machines could perform. This model is called a Turing machine in his honor.

Turing figured out how to build a version of such a machine that could be used to imitate—all at once—thirty-six of the machines the Germans used each day to create codes in World War II. Turing's "bombe," as it was called, could test out thousands and thousands of possible combinations every day. In order to narrow the options to a number that would increase the odds, Turing also invented a statistical method (called "banburismus") that gave different weights to different possible settings, which allowed the machine to rule out several thousand possible combinations each time.

This machine helped British intelligence break one of the most difficult secret code systems that had ever been invented, and because of that, it helped the Allies win World War II. The machine he developed became the ancestor of the modern computer. In addition, Turing also went on to do pioneering work in artificial intelligence and in creating computer models of different natural systems.

binary would represent? (1 × 1) + (0 × 2) + (1 × 4) + (1 × 8) = 13.

Now, what about letters? Well, they are each just assigned to certain different binary strings, like this:

$$A = 01000001$$
$$B = 01000010$$
$$...$$
$$Z = 01011010$$

At this point it is fair for you to ask: Why are we stuck with only using 1s and 0s? Because those are the only choices that electric circuits have: let the electricity through (1) or don't let it through (0). They are kind of like doors—if they are open, people can get through; if they are closed, people can't. A computer chip is made of millions of these little electrical doors called gates. Each gate is either on (open) or off (closed). If you have ever played twenty questions, you should have a good feel for how much you can accomplish with just two possible answers—yes or no.

So we can make words and strings of numbers, but the next part is that we have to be able to connect words and sentences to the real world. Here is where George Boole, a brilliant Brit, excelled. He figured out a way to show that by just adding a few operations like AND, OR, and NOT, we could slowly (and sometimes awkwardly) build a model of our language—and do it with only yes or no circuits.

You might wonder why we can just type things into a computer, or read words on a screen, instead of having type in long strings of 1s and 0s, or read long strings of binary code appearing on the screen. Well, once a program has built certain things out of binary numbers, it doesn't have to keep doing it from scratch every time. That would be like you having to slowly sound out each word letter by letter when you speak. If we opened up the code and dug down, though, we would find that this process of translating things into

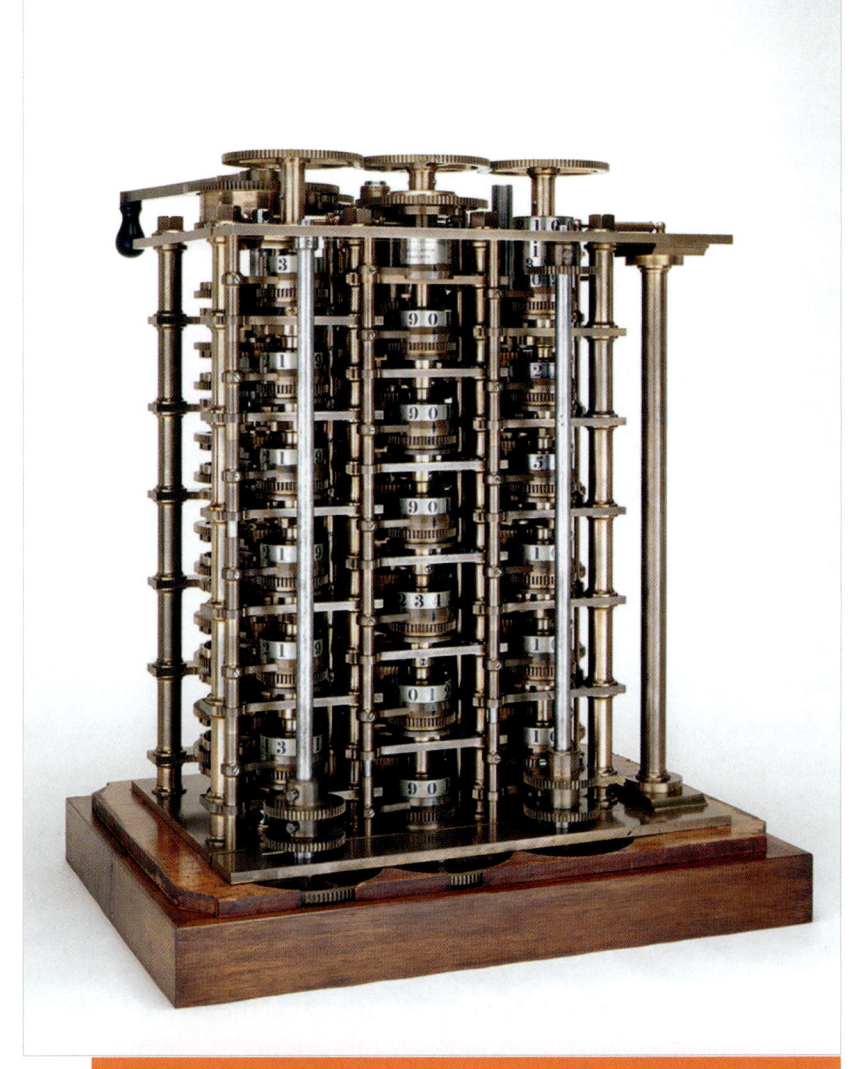

While the form of this mechanical calculator is so different, the shape of thinking that it represents still guides modern computer science.

Number and Quantity in Your Everyday Life

1s and 0s, sending them, and then translating them back to letters and numbers is what is happening at the bottom level.

Here is a basic example of how numbers play an interesting, everyday role in the Internet world—transmitting account numbers online. Imagine you are transmitting a customer identification number to stuffmart.com. Obviously you would be taking a risk to just send the number as is. Instead, what will happen is that the stuffmart.com server will send your computer a string of binary digits the same length as your customer ID. When you type your account number into the box provided, your computer will automatically translate it to binary, as we talked about. In this case, it then adds the two binary strings together in a special way and transmits the result to stuffmart.com.

Let's say your account number translates to 1001010011001. Stuffmart.com will send you a string of binary that is also 13 characters long: 1010110101101. The special way of adding them together preserves them as a binary number. If a_1 and b_1 are the same (as shown in red below), the result is a 0, and if a_n and b_n are different (as shown in blue below), the result is a 1. Don't let the little numbers and letters (subscripts) fool you. That is just a shorthand way of showing they are both in position 1, or position *whatever* (shown by the *n*), in the way that the 7 in 714 is in the same position as the 8 in 864. In this case, the result would be:

1001010011001 is the binary version of your account ID.
+ 1010110101101 is the string sent by stuffmart.com.
0011100110100 is the number that your computer sends.

If someone were to intercept that message, they would have no useful way of converting it back to the original account number because they are missing the middle number. Once the new number gets to stuffmart.com, though, they can untangle your real account number by simply adding their string a second time:

```
   0011100110100
+  1010110101101
   1001010011001
```

If you check, this is the binary number for your account. Pretty clever, right? That being said, though, more advanced, modern methods of encryption use an additional kind of math code to help make the information much more secure against various kinds of programs that would try to hack them.

Encryption

To understand this additional kind of coding, we first have to introduce a special kind of counting, called modular mathematics. It is actually something most of us use daily when talking about clocks and time. In everyday language, when we calculate, for instance, what time we'll be finished with a project, we go through however many instances of twelve might occur, but then we focus in on the amount that is left over. For example, if we start working on a project at 9:00 in the morning and expect to work for six hours, we typically say we'll be done at 3:00 in the afternoon, not 15:00. Of course, there can also be calculations in other groups, figuring out how many buses might be needed for a field trip, for instance. These calculations are also a useful tool in mathematics itself, as they provide ways to break groups of larger digits down into groups of smaller ones, making them easier to work with.

This kind of math is also part of a common Internet encryption process called the RSA public key cryptosystem. It took its name from the inventors: Ronald Rivest, Adi Shamir and Leonard Adleman. This system works by giving you some of the numbers you can use to lock it.

Before you encrypt the message, it is first turned into a number by using a standard method where the characters are replaced by binary digits. However, because working and thinking in long

In addition to her programming, Ada Lovelace also introduced key ideas about the nature and limits of artificial intelligence.

The grandmother of computer programming

Here we meet one of the great heroines of math and computer science, Augusta Ada Byron, Countess of Lovelace—better known as Ada Lovelace. Born in 1815, she was the daughter of the poet Lord Byron. Having received an education in science and math, which was uncommon for a woman at the time, she demonstrated a great talent for math, and by the age of seventeen, she had become friends with Charles Babbage, an inventor and mathematician.

Having had some success with a mechanical calculating machine called the Difference Engine, Babbage developed ideas for an even better one, known as the Analytical Engine. Though the latter never got built, Lovelace was captivated by it, and when she was asked at one point to translate an Italian-language paper about the invention, her notes ended up being longer than the original paper. It is in these notes that we see the sparks of great ideas, like adding letters and symbols to the machine's processing abilities. Lovelace also figured out how to get the machine to repeat certain steps without having to rewrite them all, a process known today as "looping." It is from these notes and ideas that she has come to be known as the grandmother of computer programming.

binary digits would make our brains melt, we'll go on to talk about this as if they were ordinary numbers that we're more familiar with.

To add the next layer of encryption, let's pretend our message is something really simple, like "hi." If we just replace each letter with its number in the alphabet (a very old kind of code), we would get the normal-looking number 89. We'll transmit it in two pieces, an 8 and a 9.

So, stuffmart.com sends me some numbers to use, which is kind of like sending an open padlock I can lock onto my box to send the message back to them. First they come up with two prime numbers. In real life these are crazy-long prime numbers, but for this example we will use 11 and 13. They multiply those together to get 143.

To understand why they do this, and how much of a difference it makes to computers that might try to hack the secret code, think of how hard it is to go backward compared to forward. How long would it take you to break 15 into its prime factors? Not long, right? Take a minute and start working on 481. It'll be a while. The prime factors of 481 are 37 and 13. As the two prime numbers used get longer, the time it would take becomes ridiculous.

OK, so they send you a random prime, for instance 7, and they send you 143. Your computer would need to do two important calculations. First, it would find 8^7 mod 143.

Here, "mod" mean "modulo." It's the short mathematical way of saying "taking the remainder." For example, 13 mod 4 = 1, because you can divide 13 by 4 a total of 3 times, and have 1 left over.

$$= 2{,}097{,}152 \text{ mod } 143 \text{ (meaning the remainder of } 2{,}097{,}152 \text{ divided } 143)$$

$$= 57$$

Then, your computer would find 9^7 mod 143.

= 4,782,969 mod 143 (so, we divide that big old number by 143 and get the remainder ...)

= 48

Your computer sends 57, and then, a fraction of a second later, it sends 48. Stuffmart.com's computer then has to untangle the numbers you sent:

57^{103} mod 143 = 8 (Take the humongous number and divide by 143 for a remainder of 8)

48^{103} mod 143 = 9 (Same thing: calculate giant number, then divide for a remainder of 9)

Then the last step is to switch the 8 and the 9 back for *h* and *i* to see your message of "hi."

There is a formula that they use to come up with the exponent of 103 that is needed to untangle the message you sent, but we won't spend time on that part here.

In August 1999, a large team of mathematicians took over thirty-five computing years to factorize a 155-digit number, and the RSA site assures us that current technology cannot factorize numbers of 230 digits. As long as the two primes are kept secret, the random number they gave you (the 7) and the mod (the 143) can be shared and still be secure.

Web page ranking

The last area to look at to see how numbers and quantity operate in the realm of computers is how search engines create different page rankings. The same kind of process has application in other areas, too, as you might expect.

To see how this works, let's work with a simple diagram that only has four or five web pages. They are connected by the network, but not all pages have links to each of the other pages. We can use arrows to show this, as in the diagram on the previous page.

When we check the diagram to see which page is the most "valuable" and should get the highest ranking, it looks like A and D would tie because both have three links coming in. That means the other pages have sort of "voted" those to be the top.

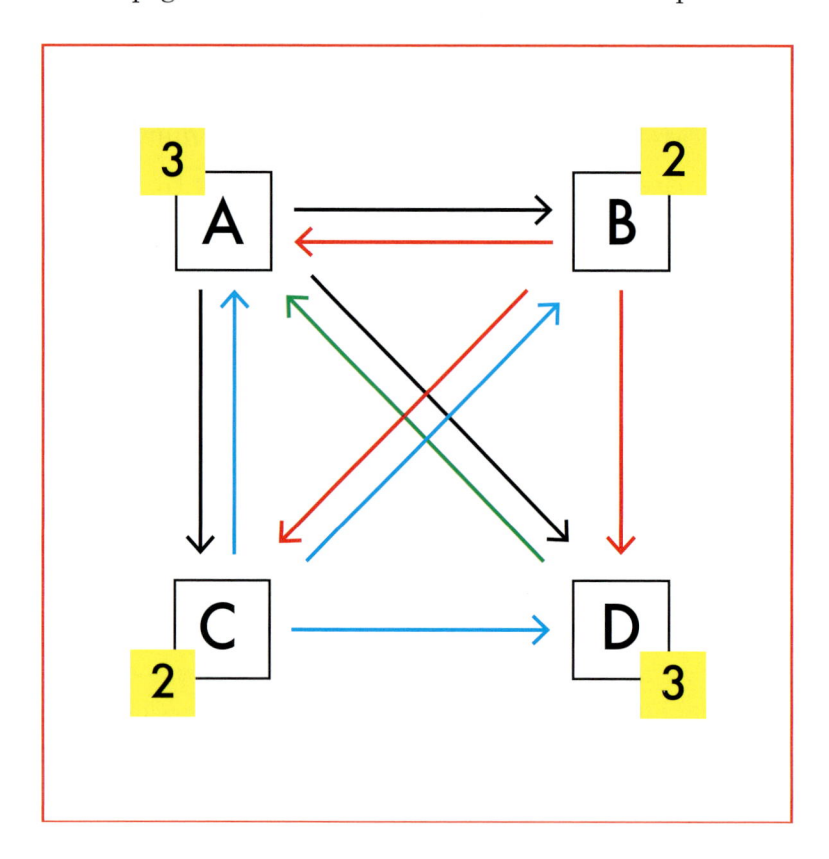

A careful consideration would suggest that not all pages are created equal, though, so why should they all get the same amount of votes? We can see that A has a link score of 3, B of 2, C of 2, and D of 3. If, however, we weight the links coming into a page based on the scores of the pages that link to it, we then see that:

58   Applying Number and Quantity to Everyday Life

$$A \text{ would get } 2 + 3 + 2 = 7$$
$$B \text{ would get } 3 + 2 = 5$$
$$C \ldots\ldots\ldots\ldots 2 + 3 = 5$$
$$D \ldots\ldots\ldots\ldots 2 + 3 + 2 = 7$$

Well, A and D are still at the top. But does the fact that a lot of links connect to pages A and D mean that they are actually better pages that should get a larger vote for the value of other pages? How do we take into account how much time a surfer might spend on the page, or how frequently it gets visited? Obviously a page that has a ton of links to it but hardly ever gets visited, or a page that people visit but then quickly leave, would not seem like as "good" a website.

The way this is resolved is to give each page a fraction of the total weight. In order for the math to work out behind the scenes, the total is set to 1, so in our simple Internet, each page would start off with one fourth of the total. During each "round" of activity, each page portions out its fraction evenly to each of the others. So, for instance, since A links to three pages, each page that A links to would get $1/4 \times 1/3 = 1/12$. Since D only links to A, A would get the total of $1/4$ from D. That means at the end of round 1:

$$A = 1/12 + 1/12 + 1/4 = 5/12$$
$$B = 1/12 + 1/12 = 2/12$$
$$C = 1/12 + 1/12 = 2/12$$
$$D = 1/12 + 1/12 + 1/12 = 3/12$$

At the start of the next round, then, A would be the top-ranked page, and would have the most votes to pass out (5/12). It would send 1/3 of those to B, 1/3 to C, and 1/3 to D. The same goes for each of the others. Again, D would send all its 3/12 to A, since that's the only one it links to.

As this process repeats over and over, the rankings will continue to sort out in similar fashion, and that determines page rank. Over a certain number of "rounds," these page rank allowances settle down to be more or less stable, unless there are links added or deleted, which can cause updates. In our simple model, it happens in as soon as ten rounds of activity. The page rankings have settled down with weights of:

$$A = 37.5 \text{ percent } (4.5/12)$$
$$B = 18.75 \text{ percent } (2.25/12)$$
$$C = 18.75 \text{ percent } (2.25/12)$$
$$D = 25.0 \text{ percent } (3/12)$$

Of course, with the real Internet there are many more variables and many more pages, so the math that goes into solving gets messier, but the idea behind the model is the same.

Numbers in Nature

Moving away from the man-made world of computers, let's turn outward to the world of nature and look at some of the ways that numbers seem be a kind of blueprint shaping the patterns of appearance and behavior in things from particles to bacteria and on out to the stars.

The growth of grass and fierce, man-eating butterflies

When we look at the way certain things grow in nature, whether that is a tree or a tortoise or a swarm of ferocious, man-eating, imaginary butterflies spreading to take over the rainforest, there is a certain magic kind of number that appears. This number goes by the name of e. That's it, just e. It is like some kind of pop star that only needs one name. Like π, e is something called a **transcendental number**. We'll get to that in a minute.

For some things, like bacteria growing in that jug of milk that has been hiding in the back of the fridge for a month longer than it should have, or a population of rabbits or butterflies, the process of exponential growth is going on continually. Since each new bacteria or butterfly is also able to produce its own little baby bacteria or pupa, the rate at which the total population goes up each week or day or hour keeps increasing. As we shrink the chunks of time thinner and thinner, we get closer and closer to a case where the growth rate is going on constantly. As that happens, the value gets closer and closer to *e*.

We can represent *e* with the following formula:

$$(1 + 1/n)^n$$

As we begin to plug numbers in, the result looks something like this:

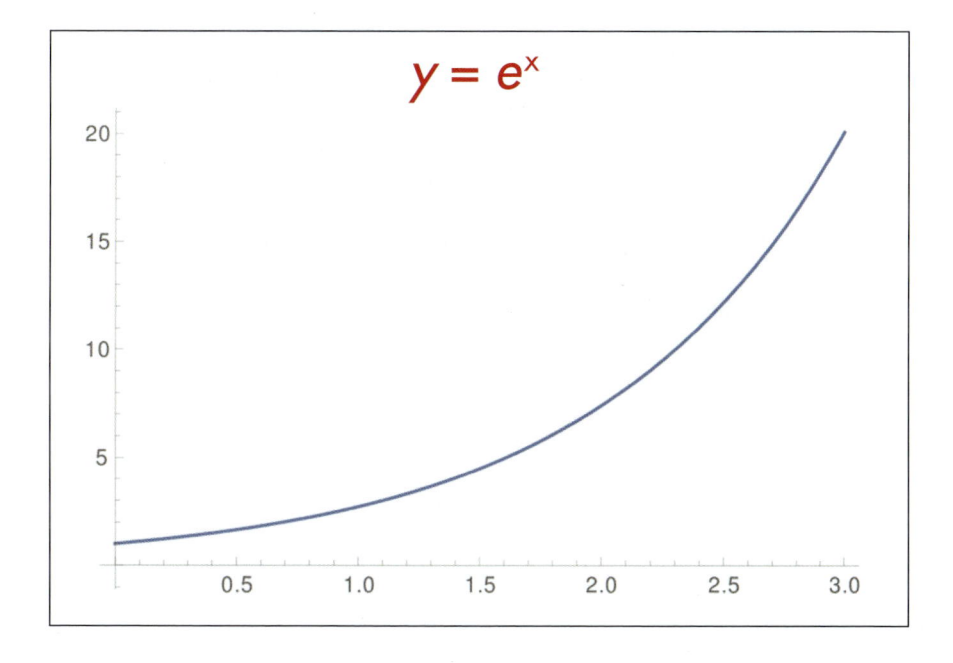

$$y = e^x$$

To get a better sense of what is going on in such a graph, consider the following list of calculations, and notice what is happening to the result as *n* gets bigger and bigger:

Number and Quantity in Your Everyday Life

Jakob Bernoulli was a member of a family filled with brilliant scientists and mathematicians.

Bernoulli and the discovery of e

While we often think of Leonhard Euler when discussing the transcendental number e, given that he was the one who began using that letter to refer to it and that he did much to advance its use, the number actually made an appearance in a different form in the work of John Napier, in 1618, as he worked to develop logarithms as a tool for changing multiplication problems into addition problems, thus greatly simplifying the multiplication of large numbers.

However, it was up to Jakob Bernoulli, the Swiss scientist and mathematician, to reveal the real applicability of e during his exploration of compound **interest** and how to calculate its value. It should come as no surprise, really, for Bernoulli came from a famed family of mathematicians and scientists. Euler was friends with the family and was tutored by Jakob's brother Johann.

In addition to his work to discover the value and importance of e, Jakob is also remembered for important, groundbreaking work in the field of probability. He first developed the law of large numbers, and one of his ideas related to number theory was the subject of an algorithm designed by Ada Lovelace for Babbage's Analytic Engine, making it part of the first published computer program.

			Difference from previous result:
$n = 1$	$(1 + 1/1)^1 = 2^1$	2	
$n = 2$	$(1 + 1/2)^2 = 1.5^2$	2.25	0.25
$n = 4$	$(1 + 1/4)^4 = 1.25^4$	2.44141	0.19141
$n = 10$	$(1 + 1/10)^{10} = 1.10^{10}$	2.59374	0.15233
$n = 25$	$(1 + 1/25)^{25} = 1.04^{25}$	2.66584	0.07210
$n = 50$	$(1 + 1/50)^{50} = 1.02^{50}$	2.69159	0.02575
$n = 100$	$(1 + 1/100)^{100} = 1.01^{100}$	2.70481	0.01322
$n = 1,000$	$(1 + 1/1,000)^{1,000} = 1.001^{1,000}$	2.71692	0.01211

As n approaches infinity (that just means as n gets bigger), the fraction inside the parentheses gets smaller and smaller (1/10 versus 1/1000), and the sum inside the parentheses gets multiplied by itself more and more times.

Continuing this series out to $n = 10,000$ gives us a value of 2.71815. Notice how close that is to the value of calculating e out to 30 decimal places:

2.71815 vs.
2.718281828459045235360287471352

Fibonacci series and bee family trees

As discussed at the beginning of the book, one of the things we use numbers for is to recognize certain patterns and arrangements in order to help us understand the world around us. One particular pattern that is interesting in this regard is the presence in different parts of nature of a sequence called a Fibonacci series. This patterned sequence of numbers acquired its name from the nickname of a famous mathematician named Leonardo of Pisa, who lived in Italy in the twelfth century. Leonardo was a great mathematician who had benefitted early on from being brought up in an environment filled with chances to learn from people trained in Hindu-Arabic systems of numbers and calculation which we discussed in the opening chapter of this book.

The pattern for this sequence of numbers is to add the two previous numbers. See how it works at the top of page 66.

It is another interesting feature of these patterns that from the number 5 on, every second number in the Fibonacci sequence is the length of the hypotenuse of a right triangle with sides that are integers (a regular whole number like 3 or 52, not something awkward like 5.17239). The length of the base of each triangle in the series is equal to the total of all three sides of the triangle before it, and the length of the shorter leg equals the Fibonacci number you just skipped over minus the shorter leg of the triangle before.

The lengths of the triangles' sides are constructed like this:

First triangle: 5, 4, and 3

Second triangle: 13, 12 (5 + 4 + 3), and 5 (8 − 3)

Third triangle: 34, 30 (13 + 12 + 5), and 16 (21 − 5)

This series continues indefinitely.

If we start with 5 as the hypotenuse, then we can probably remember that that belongs to a right triangle with measurements 3, 4, and 5. The next Fibonacci number is 8, but we skip that one. That gets us to 13, which is the length of the hypotenuse for a 5, 12, 13 right triangle. We skip 21, then have 34, for a right triangle of 16, 30, and 34. Admit it, that's a little cool, right?

Even more surprising than finding certain patterns among numbers and shapes is how often we find those same kinds of relationships in nature. It's almost like the universe was built

$$0 + 1 = 1$$
$$1 + 1 = 2$$
$$1 + 2 = 3$$
$$2 + 3 = 5$$
$$3 + 5 = 8$$
$$5 + 8 = 13$$

with math as the blueprint. Fibonacci sequences are one of the interesting mathematical patterns that seem to pop up all over the place in biological settings. There are a bunch of examples, including the bones in your finger, but we'll just look at some of them here.

In pinecones and pineapples, there is actually a double set of spirals—one twisting in a clockwise direction and one twisting counterclockwise. If you count these spirals, the two sets are found to be neighboring Fibonacci numbers.

Besides just being visible, we also find Fibonacci sequences hiding in nature, like in the family tree for a honeybee. The family tree of bees is a little different than most of us are used to thinking about. The rules for bees work as follows:

- If an egg is laid by an unmated female, the offspring is a male (drone) bee.
- If an egg is fertilized by a male, then the offspring is a female (worker) bee.

As a result, each male bee has one parent, and a female bee has two. Following that back to the grandparents, great-grandparents, etc., we find two Fibonacci series, as shown in the chart on page 67.

Number of	Drones	Worker Bees
Parents	1	2
Grandparents	2	3
Great-Grandparents	3	5
Great-Great-Grandparents	5	8
Great-Great-Great-Grandparents	8	13

This pattern can be seen on the smallest scale, where we find that the human DNA molecule is 34 angstroms long and 21 angstroms wide when we measure each full twist along its double helix spiral (like taking a ladder and twisting the top and bottom in two directions). These numbers, 34 and 21, are Fibonacci neighbors, and their ratio, 1.6190476, is almost the same as the golden ratio, 1.6180339. The pattern also occurs at the largest scale, where we see it tracing the curved arms of spiral galaxies like the Milky Way!

Fractals

Traditionally, in attempting to apply a mathematical model of understanding to different shapes in the natural world around us, we were very limited. We could wrap our brains around how things that looked like traditional geometric forms—spheres, cones, cylinders, and so on—could be analyzed using math. But a wide variety of other things were not so easy to understand. Not even if we close one eye and squint to blur the edges can we make the side of a cloud or the shape of a coastline appear smooth like the edges of a cube or a pyramid.

In 1975, however, the mathematician Benoit Mandelbrot had the first real inkling that these complex shapes do have some interesting properties. Using mathematical tools, he found that there is, in fact, a pattern in such things as the branches of a tree, a coastline, or the shapes of clouds. He gave these patterns a name—fractals—and then proceeded to try to make sense out of them and create a mathematical model that could be used to study properties of other things exhibiting the same kind of pattern.

A distinctive feature of fractals is that they exhibit the same pattern whether we zoom in to a very small scale—like the tiniest branches at the tips of trees—or zoom out to a larger scale—like the biggest branches coming off the main trunk. As you might imagine, there is a similar effect for zooming in closely to look at one little area of the edge of a cloud versus looking at the overall shape of the cloud's edge.

Once we began to understand the nature of these patterns, mathematicians and scientists were able to recognize them at work in a wide variety of other places. Examples range from the branches on a tree to the branching of the increasingly tiny air tubes inside our lungs.

How do numbers come into it? To see that, we should start with something simple and segmented, like the Koch snowflake.

In 1904, a Swedish mathematician named Niels Fabian Helge von Koch developed an example for the sake of illustrating a different mathematical point having to do with stock markets and some ideas from higher-level math. He created a process to draw increasingly complex shapes out of triangles. The process is simple: You take each straight line in a triangle, cut out the middle third of that line segment, and insert two new lines as if adding a smaller equilateral triangle to the sides of the larger one. You can see the images he came up with on the previous page.

It was only later that mathematicians like Mandelbrot realized the shape had some useful application to understanding other areas of math. When we turn to start poking around in the numbers, we will start to appreciate some of the things that made this shape (and others) puzzling and interesting.

A straightforward place to start would be to look at how we count the sides of such a shape. For each **iteration**, which means each time we perform the process, one side of the figure from the previous stage becomes four sides in the new stage. Since we start

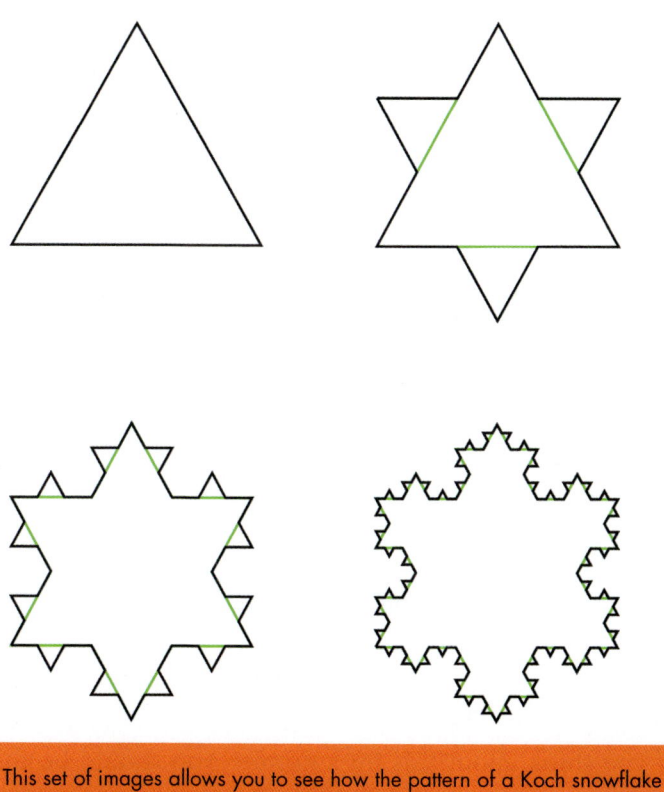

This set of images allows you to see how the pattern of a Koch snowflake keeps repeating at smaller and smaller scales.

Number and Quantity in Your Everyday Life

with three sides, the formula for the total number of sides at each stage is:

$$n = 3 \times 4^a \text{ in the } a\text{th iteration}$$

Here, ath just means whatever number represents how many times we've done the process—the second time (so a would be 2), the thirteenth time (a = 13), or the bazillionth (not a real number, but you get the idea). For iterations 0, 1, 2 and 3, the number of sides equals 3, 12, 48 and 192, in order.

Each time we go through the process, the length of a side is one-third the length of that side in the previous stage. So, if we start with an equilateral triangle that has side length x, then the length of a side in iteration a is:

$$\text{length} = x \times 3^{-a}$$

A negative exponent just means instead of 3^2, we would find $1/3^2$, or $1/9$. When we run the calculation from 0 to 3 times, the length comes out to a, $a/3$, $a/9$ and $a/27$.

Since all the sides in every iteration of the Koch snowflake are the same, the perimeter is simply the number of sides multiplied by the length of one of the sides:

$$p = n \times \text{length}$$
$$p = (3 \times 4^a)(3^{-a}x)$$

for the ath iteration. Simplifying one last step gives us:

$$p = (3x)(4/3)^a$$

If you think about this for a minute, you should be able to see that as a approaches infinity, the perimeter keeps growing longer and longer without an upper limit.

Here comes the coolest part, though: even though a fractal like this could keep going (not always in nature, due to physical limitations, but mathematically possible), reaching nearly infinite perimeter, its area would converge to a particular value. This works similarly to the way the value of e was converging in tinier and tinier steps toward a particular value. In this case, it would be eight-fifths the area of the original triangle.

A side effect of the weird difference between a limited area and an unlimited perimeter is that fractals have a dimension measurement that is in between a line (which has only one dimension, you'll remember, and that is length) and a square or other figure that has two dimensions (meaning it fills up a particular section of a plane). Each type of fractal would have a different dimension measurement. In the case of the Koch snowflake, it is considered to have 1.26185 dimensions.

As it turns out, the galaxies in the universe are also arranged in a fractal pattern, though in this case, the fractal is in three dimensions, rather than two, like the snowflake we were just examining. Oddly enough, the fractal pattern the arrangement exhibits is one that is similar to soap bubbles in a sink. It's very interesting how that resonates with the soap-bubble models used to figure out roof structures earlier.

Gravity and black holes

Sometimes math helps us find things that we didn't even know were there. In 1846, Urbain-Jean-Joseph LeVerrier was able to discover the existence of Neptune just by understanding the mathematical equations spread out all over his desk, and it was only later that anyone was able to locate it by telescope. In the same way, the mathematical work of Einstein and others has allowed us to establish the existence of black holes in the universe, even though we have not been able to see one. This is rather like being able to figure out there is a black cat running around in your yard at night because of the paw prints around your back porch—even though you've never seen or heard it.

Similarly, Einstein posited the existence of black holes based purely on the results of his mathematical model of space-time and the way space and time work with respect to gravity. We have, however, verified the mathematical model that predicts them by being able to test out a number of other things the same model predicted. That leads us to believe it is a reliable model.

Einstein is credited with the breakthrough mathematical paw prints, so to speak, that allowed us to strengthen our hypothesis that black holes exist. He published his general theory of relativity in late 1915, and only a few months later, in early 1916, another scientist, Karl Schwarzschild, found the first solution of the Einstein equations. In this case, "solution" means some values for the variables describing the curvature of space-time and the distribution of matter that satisfy the Einstein equations. That solution outlines the gravitational field surrounding a spherically-symmetric body—what we now call a black hole.

One of the curious questions at the center of understanding black holes is the question about how a star turns into a black hole. To answer that, we have to dig into the details of Einstein's model to see how his theory explains them. Einstein's model of general relativity tells us that massive objects curve space-time. The next question for most of us is: *How* do black holes affect the curvature of space-time?

Imagine a mattress made of solid rubber. If we set, say, a 15-pound (6.8 kg) bowling ball on it, there would be a dent that was larger than just the size of the bowling ball, but the bowling ball would not push all that far down into the mattress, right? Compare that to putting on the mattress a sphere of metal that was about the size of a softball but weighed 100 pounds (45 kg). It makes a smaller dent, but it sinks much farther down. Now imagine if we put a ball on the mattress that was only the size of a marble,

This is an attempt to show how space itself is bent by the power of a black hole.

but it weighed 100,000 pounds (45,360 kg). (A small spoonful of material from a neutron star can weigh 10 million tons!) That would push way, way down into the rubber mattress. So, we can think of how much steeper the sides of that dent would be than the shallow dent caused by the bowling ball. That's what we mean when we talk about the curvature being higher.

Albert Einstein's work was so insightful that some of his ideas couldn't even be tested until almost a hundred years later.

74   Applying Number and Quantity to Everyday Life

When were black holes first theorized?

In the late 1790s, John Michell (England), and shortly after, Pierre-Simon Laplace (France), used Newton's laws to argue for the existence of something like an "invisible star." They calculated the mass and size that something would need in order to exert a gravitational pull so strong that the speed needed to escape from that pull would be higher than the speed of light. Einstein's theory of general relativity (in 1915) made significant improvements on their theory and predicted the existence of black holes, but now with better reasoning behind it. It was not until 1919 that Einstein's theory of general relativity could itself be tested, so until that point, it was not quite as solid a case as we imagine it might be from our vantage point.

Part of Einstein's theory about space-time suggested that light from a distant star would be bent when it passed near enough to our sun, which bends space around it, and the effect would be for us to mistake the location of the star giving off the light. The problem was that we can only see light passing near our sun during an eclipse. So, in 1919, when there was such an eclipse, the astronomer Arthur Stanley Eddington (and others) confirmed evidence of Einstein's prediction, making Einstein a virtual celebrity overnight.

It was not until 1967 that John Wheeler, an American physicist, actually used the term "black hole" to refer to these collapsed objects.

Somewhere along that scale, the amount of steepness of the sides counts as a black hole. If you've ever tried to ride a bike up a steep hill, you know the steeper the hill, the more speed and energy you need to make it up the hill. The same is true for stars and particles, and even light. One way to characterize where to draw that line is when it is so steep that even light isn't traveling fast enough to get back up the hill, so to speak. Schwarzschild figured that value out, and so it is called the "Schwarzschild radius":

$$R_s = \frac{2G_N M}{c^2}$$

In this equation, G_N is Newton's constant (a number that always stays the same and here has to do with the pull of gravity), M is the mass of the object, and c is speed of light.

The Schwarzschild radius is, more or less, the size that a given object, with a given mass, should have in order for it to be a black hole. For instance, the Earth has mass of 6×10^{24} kilograms, so:

$$R_s = (2G_N M)/c^2$$
$$R_s = [2 \times (6.67 \times 10^{-11}) \times (6 \times 10^{24})]/(3.0 \times 10^8)^2$$
$$R_s = (8.0 \times 10^{14})/(9.0 \times 10^{16})$$
$$R_s = 0.0089 \text{ m, or } 8.9 \text{ millimeters}$$

That means for the Earth to be a black hole, its diameter would be a little shorter than the last joint of your pinky finger. Wow!

Of course, we know a planet like Earth could never be that small because there are other natural forces that work against it, but when very massive stars run out of nuclear fuel, they cool down and they collapse. Because they are so powerful, they keep crushing in on themselves until they become small enough to fit within a Schwarzschild radius—therefore, they form a black hole.

The Schwarzschild radius allows us to calculate, for example, that if we managed to concentrate the whole mass of the sun into a sphere of a few kilometers in radius, then it would be a black hole.

It is hard not to feel a little amazed that even though our math cannot make sense out of what is happening to space, time, and light inside a black hole, the numbers can still help us understand how a black hole doesn't tear the universe apart!

We've seen the ways that numbers and quantities and equations are tangled up in things around us from the scale of bacteria to galaxies and from natural objects and systems to the human-created layer of the world. We will turn now to see how math and numbers are mixed into the jobs that people do and the kinds of problems and puzzles that math helps them solve.

Galaxies show mathematical patterns—like a spiral similar to the golden ratio spiral reflected in a nautilus shell and in the Acropolis of Greece.

Number and Quantity in Your Everyday Life

✚ To a shepherd, this is not just a field with grazing sheep. There are growth curves, market values, energy conversion rates, wool volumes, and more.

THREE

Number and Quantity in Others' Everyday Lives

In the same way that it is always interesting to learn about all the different types of people in various careers, it can also be interesting to see the various ways that those people use different kinds of math to help them succeed at those jobs. In this chapter, we will just briefly introduce the different jobs, but most of the focus will be on some of the examples people give for how they use numbers and quantity—whether it's to increase profitability, keep their neighborhoods safer, or help create amazing video games that let us imagine we've headed out to conquer the universe.

Shepherds and Ranchers

Shepherds and ranchers build a career around raising livestock for different uses. Some are for sale as food and others are raised for a product, like wool or dairy. Shepherds and ranchers tend to work outdoors in all types of weather, and they must be able to hold up under the physical demands of managing and raising livestock. Oftentimes ranchers and farmers must wear a variety of hats from one day to the next, ranging from basic veterinary care of the animals and managing the business side of the ranch to maintaining the fields, fencing, and buildings or equipment needed to carry out their job.

Often, in the constant process of managing the health and growth of the sheep, the shepherd will have to use math skills to calculate conversion ratios—from metric to US standard, for instance. Metric tends to be easier to deal with because it's based on tens, so quantities are easier to calculate, but many medications will give dose amounts in fluid ounces. Sometimes they refer to the animal weight in pounds, and sometimes in kilograms.

This process of conversion is compounded when the instructions for the medication might say to give 6 cc per 100 pounds (45 kg) of weight, but the particular sheep in this case might be 80 pounds (36 kg) or 150 pounds (68 kg). If the shepherd doesn't calculate the appropriate dosage, a sheep might not get enough medicine to help it, or it might die from an overdose. Let's consider an example that could come up.

In this case, the medication comes as a powder that has to be mixed with water to make a solution. According to the directions, the standard solution uses the whole container of powder and will treat 100 sheep of 50 pounds each, giving each sheep 1.2 fluid ounces of solution. It also says if the sheep weigh within a couple pounds of 80, they should get 1.6 ounces each. The shepherd only has 65 sheep to treat, and they are within a couple pounds plus or minus of being 95 pounds each.

The first thing the shepherd will do is figure out how many ounces of the medication she should mix with the water to accurately dose her 95-pound sheep.

The general formula for extrapolation is:

$$y(x) = y_1 + \frac{x - x_1}{x_2 - x_1}(y_2 - y_1)$$

Sheep in pounds	Number of sheep	Ounces of solution per sheep
95	65	y
80	100	1.6
50	100	1.2

We'll let x represent a sheep's weight and y equal the ounces of solution per sheep. We can take the two measures given, (50, 1.2) and (80, 1.6), and plug them into the equation:

$$y(x) = 1.2 \text{ oz} + [(95 - 50)/(80 - 50)] \times (1.6 \text{ oz} - 1.2 \text{ oz})$$
$$y(x) = 1.2 \text{ oz} + (45/30) \times 0.4 \text{ oz}$$
$$y(x) = 1.8 \text{ oz.}$$

But wait. Since she only has 65 sheep instead of 100, she would adjust proportionally, otherwise she is just going to mix up more solution than she can use, and that wasted cost cuts into profits for the season.

Instead of pouring in the whole 16-ounce packet, she can just use a simple proportion: $65/100 = x/16$. Cross multiply $65 \times 16 = 1040$, then divide by 100. She will only use 10.4 ounces of the medication powder.

Police Officer

A police officer's primary role is to help protect lives and property. Uniformed police officers typically respond to calls (emergency

Meet Cindy McKenzie

Please meet Cindy McKenzie. Cindy and her husband run a farm that raises sheep.

Thanks for sharing some perspective with us, Cindy. So, what do we call your job: Farmer? Shepherd?

I like "shepherd."

What kind of training did you need to get this position and how long have you been a shepherd?

I've been doing this for five years, now. It's true that a degree is not required to get started in this industry, but if I hadn't earned a bachelor's degree in computer science and a master's degree in atmospheric science, I wouldn't have been able to succeed at it. Many people don't realize that a farmer must be a jack-of-all-trades, with diverse knowledge of many fields, including veterinary science, weather forecasting, growing and caring for plants and grasses, math, accounting, marketing, and so on.

For many people who haven't grown up around farming and ranching, it might seem like a job that would only use a little bit of basic math. Would you agree?

You would be amazed at the different kinds of math that can be involved. I'm always using simple sums, differences, multiplication and division. I use algebra, usually in the form of ratios. I often calculate averages, and sometimes I need to do linear **interpolation**—which means finding numbers or measurements for a graph or a table if it comes between numbers you already know. I calculate percentages all the time, and I need to have a rudimentary knowledge of probability. As the industry of raising sheep continues to become more and more scientific, the kinds of math that can help you be better at it and make more money go up significantly.

One of the questions we like to ask is: Is there a kind of math or kind of math problem you wish you were better at?

I wish I could do mathematics in my head more easily, like other people I know. I have taken some really advanced math classes in graduate school, and I did very well, but except for doing addition problems in my head, I need paper and pencil for even basic stuff, not to mention doing any algebra where I need to solve for x. Don't think being fast at math is the same as being good at it. Sometimes I will have to chant the times table out loud to refresh my memory, like I did when I was learning it as a kid!

Thanks to TV and movies, we often underestimate the amount of careful reasoning and psychology that are at the heart of good police work.

and non-emergency), conduct traffic stops, patrol assigned areas, obtain warrants and arrest suspects, fill out and file documentation of incidents, and prepare cases and testify in court. Detectives share many of these roles, but there are a few differences. In particular, a detective will not spend the same time on patrol but will be focused more heavily on collecting and securing evidence from crime scenes, conducting interviews with both witnesses and suspects, and other things related to her specialty.

There are any number of ways that math comes up in the work of police officers and detectives, but there is one in particular that stands out as different from things you probably think of as even being math puzzles to solve.

Let's picture a scenario where two suspects have been arrested in relation to a crime. We have the Red Queen and the Mad Hatter, who have been brought in on suspicion of being conspirators in an armed robbery that resulted in the accidental murder when the victim fell down the stairs of her house. Our detective, Alice, places the two suspects in separate detention rooms and questions them one at a time. The evidence does not seem strong enough to get a conviction in either case, so Alice is really hoping she can convince one or both of them to confess.

She lays it out for each of the two: They are both facing possible jail time for the violent assault, since there were witnesses, even if she can't prove they committed the robbery as well. For the assault, they'll probably get three years. But Alice decides to offer them a bargain. She explains that if they both confess, their jail terms could be negotiated down to ten years, instead of the twenty-five years for manslaughter. If they both deny any involvement in the armed robbery, they know they will probably only get three years for the assault charge. However, she explains that if one of them will testify against the other one, she can cut that sentence down to one year with good behavior and the other one will get the full twenty-five.

The farther satellites travel, the more mistakes occur when we send or receive information. Math helps format this data so mistakes stand out.

88 Applying Number and Quantity to Everyday Life

Error codes

When large amounts of information are being translated into binary code and sent over large distances, there are always increasing probabilities of possible corruption in the **data**. That means some of the information gets mixed up. The longer the distance and the larger the amount of information, the higher those chances are. A good example of this would be the signals transmitted back from spacecraft like the *New Horizons* craft that went to Pluto.

When scientists on Earth receive that data, they need a way to check it to find any errors resulting from the transmission of the code. Imagine trying to check and see if someone had made a mistake when they were writing down a long list of phone numbers and addresses when you don't have access to the document they were copying their information from.

One solution is to break the data into eight-digit strings and then have the computer program attach one additional digit, a ninth, at the end. If, for instance, there were an even number of 1s in the string, a 0 is added. If there is an odd number of 1s in the string, a 1 is added to the end. If the number of 1s does not match this final digit's code, then scientists would know there was an error in the transmission.

Both suspects realize that nobody confessing leaves them with just a three-year jail sentence each. They also recognize, though, that if one of them confesses, he or she will get a sweet, short sentence for cooperating with the police, and the other will go to prison for twenty-five years. If both of them cave in and confess, they end up with terms of ten years each.

The two suspects are now engaged in something called the "prisoners' dilemma," a two-person, simultaneous-move game, and each "player" has to choose between confessing and not confessing to the murder. The above scenario, with choices and possible outcomes for the players, looks like this:

		Red Queen Confess	**Red Queen** Deny
Mad Hatter	Confess	10 years, 10 years	1 year, 25 years
Mad Hatter	Deny	25 years, 1 year	3 years, 3 year

The Mad Hatter is in black, and his choices are listed down the side. The Red Queen, as you will have guessed with your own super detective powers, is in red, and her choices are listed across the top. So, for example, if the Mad Hatter denies and the Red Queen rats him out, the lower left box shows the MH gets twenty-five years and the RQ only gets one.

How will they likely solve it? Well, if the MH confesses, he gets either ten years or one. If he denies, he gets either twenty-five years or three. In this set up, confessing looks like a better choice for him. From the RQ's point of view, it looks very similar: in the confess column she ends up with ten years or one, while the deny column gets three or twenty-five (that's a lot of years for her to be away from Wonderland).

This is a very simple version of a field called game theory, and the outcomes will also depend, realistically, on each suspect's estimate of how likely the other person is to be loyal or to sell them out. The really interesting thing to see about the way Alice uses math here is that she doesn't end up having to do calculations afterward, she does her math beforehand—in the way she sets up the offer.

Interestingly, this same kind of math comes up in all kinds of other areas from **economics** to military strategy.

Video Game Designer

There is not really one particular job of game designer, as you might have imagined. Making a video game involves a wide mix of different workers with very different skills and specialties. Examples of these different roles might be game design, programming, writing, physics, and art. If they work at a smaller studio, sometimes developers might have more than just one role, but the idea of putting on different hats with different skills still applies.

If we tried to cover all the different roles that go into video game development, we could easily find a whole smorgasbord of types of problems that come up and that we could use math to help us solve. For now, let's just focus in on one particular kind and see an interesting kind of problem they try to solve to make the game more realistic, and how they use math as a shortcut to solving that kind of problem.

Video games don't have all the same limits as the real world. In order to make them seem real, physics programmers have to design and write the code to imitate natural laws—like those for how gravity affects things. Working from the vision of the game designers, the physics programmers build up the set of rules that govern how all the different objects in a game will act and interact—like what should happen when a ship slams into an asteroid.

One of the particularly advanced ways that a video game designer uses math is in working to make the game more realistic.

Meet John Pritchett

Meet John Pritchett, a video game designer.

Thanks for offering to talk with us about your job, John. Can you start out by just telling us your actual title and how many years you've been in this position?

I am a Physics Programmer at Cloud Imperium Games, and I've been here about two years.

What kind of training did you need to get this position?

I have a Bachelor of Science in Engineering Physics with a focus on Robotics. In addition to that, I also had some experience working on programming for a few other video games before I came to work here.

What are some different ways and kinds of math that you use in your job? That is, what do they help you accomplish?

I work on a space simulation game called *Star Citizen*. My primary job is to create and maintain the spaceship flight simulation, so I use math to define the flight mechanics of spaceships as well as to create flight control systems for flying them.

For flight mechanics, it's mostly a kind of 3-D math which includes position, velocity and acceleration. An example would be converting the force generated by thrusters mounted on a spaceship and relating that to the ship's center of mass to figure out the direction the ship would move, the direction it would point toward, and how fast it would do those things. If you've ever seen a car skid or slide around on ice, you can appreciate that the direction the wheels are pushing and the direction the car actually moves do not always match up.

For flight control, I use a special kind of feedback controller to convert a pilot's control inputs into spaceship movement, very much like a control system of a modern aircraft or a real spacecraft. Though a video game can simplify some of the ways things behave inside the game, I work to make it more inexact to make it more like real flight control. For example, a thruster might become damaged and bent, causing it to thrust in a slightly different direction than the system tells it to. The system makes no assumptions about whether any action has the intended result, and it uses feedback, constantly measuring the velocity and other properties of the ship, to correct for errors between what it tried to do and what it actually did.

Some of the math involved is very complex and time consuming, so the designers can use a shortcut. (Remember, that is a lot of what math is about.) By using a procedure called a **Laplace transform**, a designer can take one math problem and convert it into a different kind in order to make it more manageable, simplify it, or shrink it down, then convert it back into the original kind and let it run. This is a common tool in control system design.

That might sound confusing, so let's dig into a simple kind of example to show what is happening. A typical use of a Laplace transform is to solve a **differential equation** by converting it into an algebraic equation. In the video game, it involves a bunch of pieces at the same time, so for now, let's consider an example that is just looking at one piece at a time. In nuclear physics, radioactive decay follows the equation:

$$dN(t)/dt = -AN(t)$$

where $N(t)$ is an equation about something happening over time and $dN(t)/dt$ is the way to calculate how fast or slow it is changing over time. A is the amount of the radioactive substance. Here we are talking about not just slowing down or speeding up, but how fast the decay itself happens (when radioactive material pings off pieces of itself like little random cannonballs punching deadly little holes through whatever it hits). Think about how many times a car speeds up or slows down during a trip. It doesn't just go the same speed the whole time. Consider that rolling to a stop and slamming on the brakes both bring the car speed down to 0 miles per hour, but slamming on the brakes bring you down to 0 much faster, right? So in working on puzzles like video games or how fast plutonium is using up its energy, we try to solve for the equation $N(t)$.

First, move the terms to one side of the equation:

$$(dN(t)/dt) + (AN(t)) = 0$$

The more complex the math is that goes into building a video game, the more the user can temporarily forget it is just a game.

Number and Quantity in Others' Everyday Lives   95

Pierre-Simon Laplace's patience and his ability to stay focused on one project are as inspiring as the ideas he came up with.

Laplace and the solar system

Pierre Simon Laplace was a French thinker born in 1749. He is a great example of how dedication to math and science can lift someone up in the society of his time. Laplace was not only an astounding and insightful mathematician, making huge contributions to calculus and statistics, but also an accomplished scientist—including work in thermodynamics and the gravitational effect of moons on their planets. He was also the first to suggest that the solar system had formed from a nebula of hot, swirling gases, as well as the first to theorize the existence of gravitation collapse causing something similar to black holes.

Laplace was one of the teachers at the Royal Artillery Corps who graded the tests of young Napoleon Bonaparte, and he later held a position in the French Senate under Napoleon's rule. Laplace was made a count in 1806, and he became a marquis a few years later, in 1817.

Laplace had incredible focus on one problem in a field where many divide their attentions among a variety of compelling problems—and his masterwork on giving a rigorous mathematical analysis of the solar system, *Celestial Mechanics*, was published in parts over the course of twenty-six years. Even his groundbreaking work in statistics and probability was done only because he needed it to solve some of the puzzles of the solar system.

Solving this as is involves some hard and slow math procedures, so one way to tackle it is to look on a table of Laplace transforms to see how to convert it to an algebra problem that is easier and faster to tinker with. The table would show us that the Laplace of $[dN(t)/dt] = [sN_s(s) - N(0)]$ and the Laplace of $[A \times N] = A \times N_s(s)$, so the algebraic version of the equation is:

$$sN_s(s) - N(0) + AN_s(s) = 0$$

No cause for panic here. There is no test on this stuff at the end of the chapter. If it helps, imagine fixing the carburetor on your car. Sometimes it's way easier to take it off and work on it, then hook it back up to run, right? Same idea here. The Laplace transform is just a kind of tool that shows what it will look like if we went the long way and took the original equation apart one piece at a time.

Since we now have a simple algebra equation, we can tinker with it to fix it the way we want. Here we use a little algebra skill to get $N_s(s)$ on one side by itself:

$$sN_s(s) - N(0) + AN_s(s) = 0$$

Add $N(0)$ to both sides:

$$sN_s(s) + AN_s(s) = N(0)$$

Factor the $N_s(s)$ from both terms on left side:

$$N_s(s)(s + A) = N(0)$$

Divide both sides by $(s + A)$:

$$N_s(s) = N(0)/(s+A)$$

98 Applying Number and Quantity to Everyday Life

Then the game designer would change it back going the other direction by taking the inverse Laplace transform. That means we switch it back from algebra into the more complex kind of math. As a result, the programmers end up with an equation for $N(t)$ as:

$$N(t) = N(0)^{-At}$$

This is the equation for radioactive decay. So now they could plug a time value in, like $t = 300$ years, and see whether, at the point where it has been decaying for 300 years, the plutonium is slowing down like rolling to a stop or slowing down like slamming on the brakes. In video games, programmers are not likely to be creating a simulation of plutonium giving off radiation. Instead, they might try to have the computer imitate, say, how fast your spaceship is spinning if an asteroid smashes against it. It would be messier than the example we just did, but it works the same way.

Psychologist

In comparison to the programmer working with wiring and circuits, a psychologist works with the squishy gray blob stuffed into your skull. Psychology works to understand and explain thoughts, emotions, and behavior both alone and in our different relationships with other people. Psychologists help others change the programming they have running so that it works better. They use techniques such as observation, experiment, and **assessment** in order to build and test theories to explain and improve the way we think and the different actions that follow from that.

Often, psychologists collect information and evaluate the observed behaviors by setting up controlled laboratory experiments, by working with clients at their office, and by studying different situations out in the real world. While there are a lot of different areas that psychologists can specialize in, they all share a common puzzle, and that is that in order to compare things and understand them, the psychologists have to figure out a way to turn different feelings or attitudes into measurements. If they could just hold up a color chart or a ruler to compare your anger with someone else's, their job would be easier.

One clever example of doing this was to set up a rigged board game where the rules and the beginning situation created a huge wealth advantage for one of the players, and then they could measure different things like how often the wealthy player made jokes about how good they were or how much they hogged the snacks that were sitting on the edge of the table.

Another great example was to set up a situation where a person was standing at a crosswalk waiting to cross. There was not a stoplight here, but there was a sign to yield to pedestrians. The research team wanted to test to see if wealthier drivers tended to zip right through the crosswalk, ignoring the sign, the law, and the person trying to cross. They were able to collect data by counting how many cars in each price range they saw and how many of the cars in each price range did or did not stop to let the pedestrian cross.

They tracked cars at different crosswalks and different times of day. They classified the cars based on cost tiers. The next step was to analyze the numbers that came up. Often, just because something seems like an important difference doesn't mean it really is. In the same way, sometimes things that seemed little or unimportant can turn out to be a big part of the puzzle for why people think or act a certain way.

So, once they have collected their data from observations, psychologists need to perform a few mathematical operations to see if their idea is promising enough to dig deeper and put more work into the research project.

Let us say that the researchers counted, and all of the cars in the lowest cost range stopped for the pedestrian. As the price of the vehicle rose, the percentage of people who zipped right through, breaking the law, did too. For the highest cost range, 50 percent of drivers broke the law. Let's see how they would turn their measurements into a model that would allow them to see if the two things were really related, or if they just seemed like it, and to see if it is worth doing more research to get a more complete model that explains even better.

There is a long way to do this and a short way. Let's just compare them, and you can see for yourself how a little extra math can make the short way more appealing.

The Sum of Least Squares method compares many straight lines to find the one that best fits the data:

1. Plot each data point onto a graph.

2. Draw a possible line through the group, and calculate the distance between each point and the line you've drawn.

3. Square each of the distances (to get rid of the negative values).

4. Calculate the sum of the squares.

5. Then go back and do steps 2–4 again for a different possible line. Repeat several times.

Whichever line has the lowest sum of squares (from step 4) is named Miss Line of Best Fit, gets a special first-place ribbon and tiara, and participates in special activities like parades and grand openings until she is replaced by a better line from a better experiment.

Calculating the sum of squares for each possible line by hand would be very time consuming. But, with math, there is a faster way!

The formula for a straight line is:

$$y = a + bx$$

Stop for a minute, next time you're in a car, and consider all the numbers involved: speed, reaction times, volume inside, and so on.

Applying Number and Quantity to Everyday Life

Number and Quantity in Others' Everyday Lives   103

For our purposes:

y = Percentage of drivers who didn't stop.
x = Price category of car, 1 being lowest, 5 being highest.
a = The constant (value of y when x = 0).
b = The slope of the line.

The formula used for calculating the line of best fit is

$$b = (n\Sigma xy - \Sigma x \Sigma y)/(n\Sigma x^2 - (\Sigma x)^2)$$

$$a = (\Sigma y - b\Sigma x)/n$$

where n = number of data points selected, and Σ is just a Greek letter (sigma) we use to mean "the sum of."

x	y	xy	x^2
1	1	1	1
2	15	30	4
3	24	72	9
4	38	152	16
5	50	250	25
Σ = 15	Σ = 128	Σ = 505	Σ = 55

Now we can plug those data numbers into the equation, and that gives us:

$$b = (5 \times 505 - 15 \times 128)/(5 \times 55 - 15^2)$$
$$= (2525 - 1920)/(275 - 225)$$
$$= 605/50$$
$$= 12.1$$

$$a = (128 - 12.1 \times 15)/5$$
$$= (128 - 181.5)/5$$
$$= -53.5/5$$
$$= -10.7$$

This gives us a final line of:

$$y = -10.7 + 12.1x$$

When they use the new equation to draw a line on the graph with the data points, psychologists are able to see that there appears to be a positive correlation—that is, the two variables are related. That is evidence that the one variable causes a change in the other one. In this case, having a more expensive car seems to cause the increase in mean driving. This is not proof, though. This is an early part of the process, but now the researchers would know that the project was worth digging in to explore more carefully.

Banking Manager

Typical responsibilities of a bank manager include managing a staff and working with some of the clients directly. These both include different things, from hiring and training new employees to reaching out through marketing to attract new customers to the bank. In terms of the banking operations themselves, the bank

Meet Robyn Applegarth

Here's our chance to hear about math in the everyday life of a banker, Robyn Applegarth.

Can you start out by just telling us your actual title and how many years you've been in this position?

I'm a Bank Branch Manager, and I've been in my current position for one year, but about ten years total in banking. It might surprise you, but my college degree is actually in English.

How would you describe the majority of math you have to do every day for your job?

A lot of the math I use in my job is fairly basic math, such as adding, subtracting, multiplying and dividing. This could be to help my customer budget so they aren't overdrawing their account as much, thus saving them money, or it could be helping customers figure out how to save money by potentially refinancing their loan with us. I love this part about my job because it really helps my customers in their financial lives.

You must have numbers bouncing around in your head all the time, don't you?

Well, we use lots of calculators throughout the day. I couldn't function very well in my job if I didn't have all the different financial calculators available to help save time and double-check for mistakes. If you don't understand what the math is doing, though, it doesn't do you any good to get the right answer to a problem if you've typed in the wrong problem or the wrong kind of problem.

Is there one kind of math that stands out as being a little more interesting?

One of my favorite calculations is a rather simple one. It shows how much contributing to your employer's retirement plan now will pay off in the future. Many of my younger customers don't realize the value in contributing to their retirement now, and I hear customers say all the time that they won't ever be able to afford to retire. With this calculator, I am able to show them the impact that their contributions really make over the years. The difference between if they started contributing now as opposed to waiting another ten years to start is generally thousands of dollars. It's really an eye opener.

What do you wish you were better at, when it comes to math?

Overall, I wish I were better at mental math and didn't need to rely so much on calculators, but it's important to be accurate in my position.

manager is convincing people to save and invest their money in their bank, and then, on the other side, offering to lend some of that money you saved to other people to help them grow their business, repair and improve their house, or perhaps buy a new car.

Each part of the manager's job has different kinds of math in it. That is probably little surprise—money practically *is* just a set of numbered pieces of paper; of course that is going to involve a lot of arithmetic and calculation.

To illustrate one of the most important things a bank does, let's consider the case of you, a client, who wants to buy a new car, but you don't have enough money saved up. You need the car now, not in the future when you've saved some money. To solve that problem, you could go to the bank for a loan allowing you to get the car now, and pay for it a little bit at a time over the next few years.

Bailey Savings and Loan lends you the money to buy your first car. Let's say the car costs $23,000 and you are able to put $3,000 down and borrow the rest. If the bank lends you the money at 6 percent annual interest, then they would generate a calculation that shows how much your car payment needs to be each month in order to pay off the loan at the end of five years.

	Amount Outstanding	Principal	Interest	Total Payment
1	$20,000.00	$286.66	$100.00	$386.66
2	$19,713.34	$288.09	$98.57	$386.66
3	$19,425.25	$289.53	$97.13	$386.66
[...]				
60	$384.73	$384.73	$1.93	$386.66
Total		$20,000.00	$3,199.60	$23,199.60

You should notice a couple things if you examine those calculations. For instance, when you make a payment each month, only part of it applies to the amount you borrowed, called the **principal**. The other part is paying the interest—kind of like paying the bank a certain amount because they essentially allowed you to rent the money from them.

The second valuable thing to notice is that over time, you ended up paying more than a simple 6 percent of $20,000. If you are curious about how much, just divide total interest by total principal.

Transportation Planner

Transportation planners develop the transportation plans and programs for an area (like a city or a county). They work with teams from other government and corporate organizations to determine the most important transportation needs and issues, measure the impact of the current transportation system, and forecast future transportation patterns. For example, as growth in a downtown area of the city creates an increase in jobs, the need for public transit systems to get workers to those jobs increases. Transportation planners work to develop possible ways to make sure the roads and resources will continue to work smoothly as those changes begin to unfold.

Some of the ways they use math in their job include: designing research techniques, modeling traffic flows, analyzing and interpreting data, and making presentations about the options they've come up with.

To illustrate this, let's look at an example, called the Braess paradox. To travel from A to B, the travelers either cross Bridge 1 and then travel along Route L or travel along Route R and then cross Bridge 2. The bridges both clog traffic such that the time (in minutes) to cross them is the number of cars per hour divided by one hundred. The highway sections always take fifteen minutes. When the rush hour traffic reaches a state of equilibrium, the time

This photo of nighttime traffic allowed the camera to absorb light longer, so the pattern of movement seems to be happening all at once.

Number and Quantity in Others' Everyday Lives

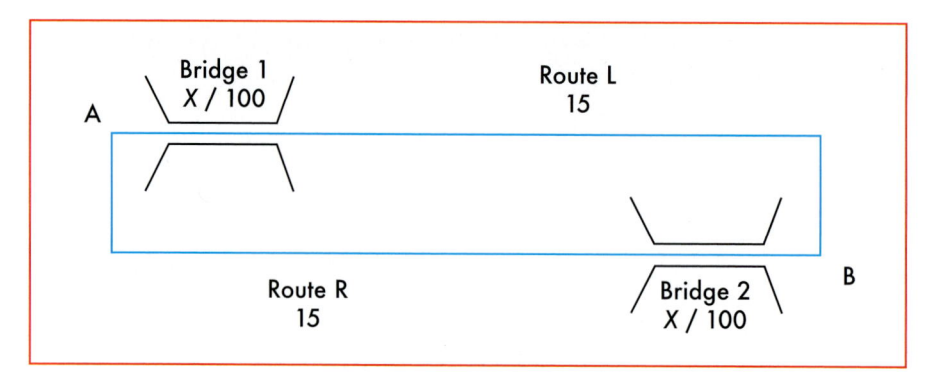

[Diagram: Route from A to B. Top path: Bridge 1 (X/100) then Route L (15). Bottom path: Route R (15) then Bridge 2 (X/100).]

it takes on both routes is the same. Let's say that:

L_t = The time it takes to travel Route L

R_t = The time it takes to travel Route R

L_v = The volume of traffic on Route L

R_v = The volume of traffic on Route R

Since they are in equilibrium,

$$L_t = R_t$$

This means that:

$$L_v/100 + 15 = 15 + R_v/100$$

If there are 1000 cars trying to get from A to B, then we know that:

$$L_v + R_v = 1000$$

And thus we can see that:

$$L_v = R_v = 500$$

112   Applying Number and Quantity to Everyday Life

If we substitute the results into our first equation we can figure out that the travel time for all drivers (on both Route L and Route R) is twenty minutes: fifteen minutes of normal travel plus five minutes of congestion to get over the bridge. The transportation planners want to improve on that number by building a high-speed option, Route C, that only takes 7.5 minutes to travel on no matter what volume of traffic is involved.

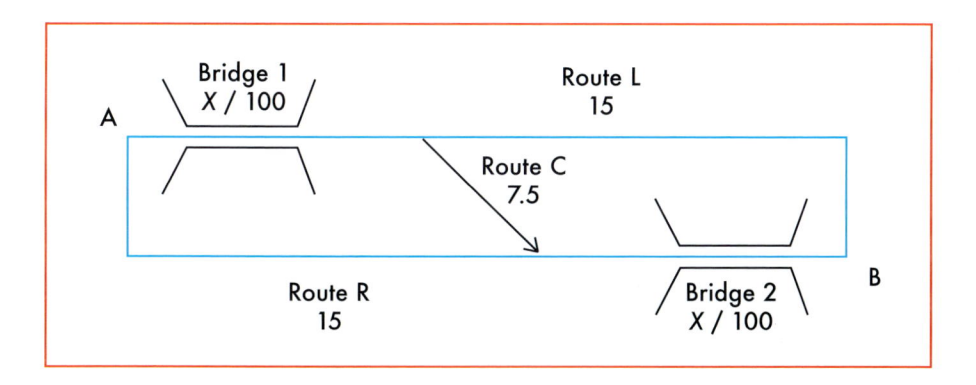

```
A — Bridge 1 (X/100) — Route L (15) — B
A — Route R (15) — Bridge 2 (X/100) — B
Route C (7.5) connects the two middle points
```

Let's label the traffic that passes over Bridge 1, then the new link, and then over Bridge 2 as C. The travel times for the three routes (L_t, R_t, C_t) in their new state of equilibrium are:

$$L_t = R_t = C_t$$

To see the problem created, we need to look at what that will mean for the travel times. Everyone leaves from point A either on Route L or Route R road, so the volume passing over Bridge 1 is all cars planning to continue on L, plus all the ones planning to take C.

In our first piece for Route L:

$(L_v + C_v)/100$ is the time spent waiting for the bridge

We add 15 minutes to travel the rest of the way.

Number and Quantity in Others' Everyday Lives

The same thing happens to those who take Route R and are joined in by people from C. They drive fifteen minutes, then wait for the second bridge. Then we have the people who take the short cut—they get both bridges:

$$(L_v+C_v)/100 + (C_v+R_v)/100$$
is the time spent waiting for both bridges

We add 7.5 minutes for the shortcut.

To see how they balance out, we put all three together:

$$(L_v+C_v)/100 + 15 = (L_v+C_v)/100 + 7.5 + (C_v+R_v)/100$$
$$= 15 + (C_v+R_v)/100$$
$$L_v + R_v + C_v = 1000$$

From this we can crunch the pieces around to show that:

$$L_v = R_v = 250 \text{ and}$$
$$C_v = 500$$

This means it now takes 7.5 minutes to cross each bridge. That's the paradox. Something that seems to shorten commute times actually causes the travel time for each route to ratchet up to 22.5 minutes. Therefore, building the new road has merely made the daily commute worse, not better.

This is a great example because it allows us to see so clearly how something many of us would be sure would cause an improvement turns out to make a situation worse. Think of how doing the math problem allowed the planner to test out the idea without spending millions of dollars of taxpayer money only to have citizens angry at both worse traffic *and* you spending their tax money on something so useless.

CONCLUSION

Well? What did you think? Were you surprised at some of the places math showed up? Perhaps an even better question is: Were you able to see that the math equations were helping us to see or do cool things that would have been much harder without them—like transmitting computer information securely, or understanding why we aren't flung off into space? Of course, I hope along the way you also saw that the math was a little bit interesting, just in the way that any riddle or crossword puzzle can be interesting.

One of the most interesting things about mathematics, and there have been some famous mathematicians who have written back and forth about this over the years, is that even when we explore something that doesn't seem to be useful for anything, it usually turns out later to be incredibly useful! Sometimes it helps solve a problem in the real world, and sometimes it helps solve a problem about figuring out how other math puzzles can help us solve a problem in the real world. In between those are the cases where we develop a math tool or strategy to help us figure a way around one problem, and it also turns out to be good for finding a shortcut through or around other problems that we hadn't thought about yet. Let's wrap up this first adventure with one of those examples.

Bridges of Konigsberg

This is a great example of how curiosity and the effort to find a way to solve certain kinds of puzzles we bump into in our normal lives can end up leading to a new field of mathematics. As a result, it also develops a set of tools to solve other related puzzles and problems as well, ranging from how to make sure your company's delivery trucks get to all their drop-off addresses on time to how to make the Internet work faster.

The people who lived in old Konigsberg, Germany (now part of Kaliningrad, Russia), wondered and debated about whether it was possible to take a walk around the heart of Konigsberg in such a way that you would cross all seven bridges, but only once each. It helps to see a model of the islands in the river that ran through the heart of the city and the bridges that connected those small islands. So, on page 117 we see a map of the city with the river colored in blue and the bridges numbered 1 through 7.

Try it. Sketch the map of the city on a sheet of paper and try to plan your trip with a pencil in such a way that you trace over each bridge once and only once and you complete the "plan" without lifting your pencil off the page.

The process of working out how to systematically prove which kinds of arrangements of pathways can be traveled without repeating has helped to generate new fields of mathematics, like topology—which is the study of how certain properties stay the same when different kinds of shapes are twisted or stretched. Examples of topology at work include the Internet and traffic jams.

So, if you like thinking about how to find simple ways to shortcut around long, slow problems, then you will understand why math might be the kind of thing people could get excited about. It has been called the greatest game ever invented.

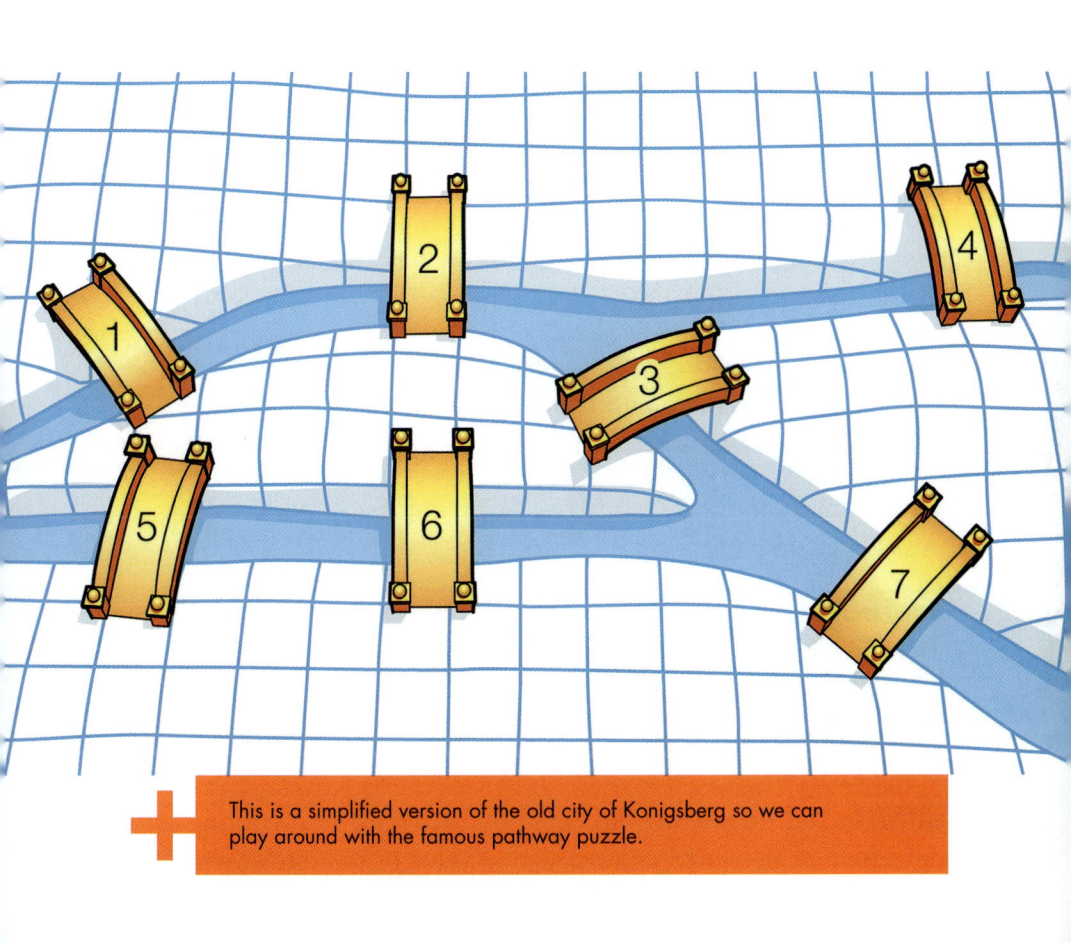

This is a simplified version of the old city of Konigsberg so we can play around with the famous pathway puzzle.

Conclusion 117

Glossary

altitude Height of an object above the ground or above sea level.

assessment A judgment or measurement of something.

axiom A statement or proposition that is accepted as being true, often based on its definition or how hard it is to imagine the result if it were not true.

baseline The starting point, often a measurement of things before making changes and measuring how much the change caused the measurement to move up or down.

binary Something that only has two members or two possible values.

coefficient The number placed as a multiplier in front of some variable, as in $5x$, where 5 is the coefficient of the variable x.

data Information gathered about an object, usually in the form of numbers.

differential equation An equation involving a function's derivative. A derivative is a calculated value of the rate at which a function is changing.

drag The amount of force pulling on something.

economics The study of the production and distribution of goods and services.

fluid A substance that has no fixed shape, adjusts easily to outside pressure, and is capable of flowing.

geometry The branch of math concerned with points, lines, planes and their relations and connections.

interest Money paid at a regular rate in exchange for a loan of some money.

interpolation Adding a new piece in between two others, like a new number on a graphed line.

irregular A geometrical shape or object in which not all sides are the same length and not all angles are congruent.

iteration A repetition of the procedure, such as repeating the same steps.

joule A standard measurement of work or energy.

kinetic From or relating to things moving.

Laplace transform A process for mapping a kind of problem from a time problem to a frequency problem in order to simplify it.

mass Roughly, the amount of atoms of a substance in an object, measured by how hard it is to move the object.

optimize To make something as good or as valuable as possible.

paradigm The best model of something; the version to measure others against.

paradox A statement or proposition with two parts that seem to contradict each other.

perpendicular Describing two objects at a ninety degree angle from each other.

plane A straight, flat surface; a geometric object defined by a surface passing through any three points not all on the same line.

potential Stored up or unused capacity; amount of work or energy that is available but not yet expended.

pressure The force exerted against an object.

principal In finance, this is the original portion of the deposit or the loan, before adding the interest earned (savings) or paid (on a loan).

probability How likely it is that a given event will occur; the mathematical study of the likelihood of different possible outcomes.

proposition A statement that contains some claim that could be true or false.

ratio A comparison between two different amounts.

transcendental number Numbers like π and e that cannot be the root of an algebraic polynomial that has integers for coefficients.

velocity The speed of something in a given direction.

Further Reading

Books

Ellenberg, Jordan. *How Not to Be Wrong: The Power of Mathematical Thinking.* New York: Penguin, 2014.

Jackson, Tom, ed. *Mathematics: An Illustrated History of Numbers.* New York: Shelter Harbor Press, 2012.

Pickover, Clifford A. *The Math Book: From Pythagoras to the 57th Dimension, 250 Milestones in the History of Mathematics.* New York: Sterling, 2009.

Sardar, Ziauddin, Jerry Ravetz, and Borin Van Loon. *Introducing Mathematics.* Cambridge, UK: Icon Books, 1999.

Stewart, Ian. *Nature's Numbers: The Unreal Reality of Mathematics.* New York: Basic Books, 1995.

Websites

Agnes Scott College
www.agnesscott.edu/lriddle/women/chronol.htm

This site has a great collection of biographies of important women mathematicians down through history.

American Mathematical Society
www.ams.org/mathimagery

While being a useful resource in general, the collection of examples connecting math and art is particularly interesting (and rare). Under the heading of Math Samplings, they also have a nice list of additional websites worth exploring.

Coursera
www.coursera.org

Here is a site with a number of different online courses at different levels of mathematical ability. It's a great, low-stress opportunity to dig into subjects that you find a little more interesting but might not have gotten a chance to work on in regular math classes.

Drexel University Math Forum
mathforum.org/dr.math

This site offers a variety of basic explanations of concepts and examples. One of its strengths is that the topics are broken down by subjects and by level in school (elementary, middle, high school, college and beyond). They do a helpful job of working through example problems and also allow you to search by topic.

James Sousa Tutorials
www.mathispower4u.com

This website provides a number of video examples of walking through problems in geometry, algebra, trigonometry, and calculus. It is not the best introduction to topics, but it is a great way to review material that is a little rusty or not quite clear in your head.

National Council of Teachers of Mathematics
illuminations.nctm.org

While this site might seem like it is just for teachers, think about it: teaching yourself would follow the same principles. The lessons and the interactive tools (like a graphing tool for functions) can be a useful way to think about concepts and principles of math. The brain teasers are also a fun way to play with ideas.

Plus Magazine
plus.maths.org/content

This online magazine is run under the Millennium Mathematics Project at Cambridge University. It offers interesting short articles on interesting problems in math (and science). The problems are complex, but they are broken down in a way that is easy to follow.

Bibliography

Barnett, Raymond A, Michael R. Ziegler, and Karl E. Byleen. *College Algebra with Trigonometry*. New York: McGraw-Hill, 2001.

Bellos, Alex. *The Grapes of Math*. New York: Simon and Schuster, 2014.

Berlinghoff, William P., Kerry E. Grant, and Dale Skrien. *A Mathematical Sampler: Topics for Liberal Arts*. Lanham, MD: Ardsley House Publishers, 2001.

Byers, William. *How Mathematicians Think: Using Ambiguity, Contradiction, and Paradox to Create Mathematics*. Princeton, NJ: Princeton University Press, 2007.

Cady, Brendan, ed. *For All Practical Purposes: Mathematical Literacy in Today's World*. New York: W. H. Freeman and Company, 2006.

Charles, Randall I., *Algebra I: Common Core*. New York: Pearson, 2010.

Clawson, Calvin C. *Mathematical Mysteries: The Beauty and Magic of Numbers*. New York: Plenum Press, 1996.

Devlin, Keith. *Mathematics: The Science of Patterns: The Search for Order in Life, Mind and the Universe*. New York: Henry Holt, 2003.

Ellenberg, Jordan. *How Not to Be Wrong: The Power of Mathematical Thinking*. New York: Penguin, 2014.

Frenkel, Edward. *Love & Math: The Heart of Hidden Reality*. New York: Basic Books, 2013.

Gordon, John N., Ralph V. McGrew, and Raymond A. Serway. *Physics for Scientists and Engineers*. Independence, KY: Cengage Learning, 2005.

Jackson, Tom, editor. *Mathematics: An Illustrated History of Numbers*. New York: Shelter Harbor Press, 2012.

Lehrman, Robert L. *Physics The Easy Way*. New York: Barron's Educational Series, Inc., 1990.

Mankiewicz, Richard. *The Story of Mathematics*. Princeton, NJ: Princeton University Press, 2000.

McLeish, John. *Numbers: The History of Numbers and How They Shape Our Lives*. New York: Fawcett Columbine, 1991.

Paulos, John Allen. *Innumeracy: Mathematical Illiteracy and Its Consequences*. New York: Hill and Wang, 1988.

———. *Beyond Numeracy: Ruminations of a Numbers Man*. New York: Alfred A. Knopf, 1991.

Peterson, Ivars. *Islands of Truth: A Mathematical Mystery Cruise*. New York: W. H. Freeman and Company, 1990.

———. *The Mathematical Tourist: Snapshots of Modern Mathematics*. New York: W. H. Freeman and Company, 1988.

Pickover, Clifford A. *The Math Book: From Pythagoras to the 57th Dimension, 250 Milestones in the History of Mathematics*. New York: Sterling, 2009.

Rooney, Anne. *The Story of Mathematics: From Creating the Pyramids to Exploring Infinity*. London: Arcturus, 2015.

Sardar, Ziauddin, Jerry Ravetz, and Borin Van Loon. *Introducing Mathematics*. Cambridge, UK: Icon Books, 1999.

Stewart, Ian. *Nature's Numbers: The Unreal Reality of Mathematics*. New York: Basic Books, 1995.

Stewart, James, Lothar Redlin, and Saleem Watson. *Elementary Functions*. Ohio: Indepdndence, KY: Cengage Learning, 2011.

Strogatz, Steven. *The Joy of X: A Guided Tour of Math, from One to Infinity*. New York: Houghton-Mifflin, 2012.

Index

Page numbers in **boldface** are illustrations. Entries in **boldface** are glossary terms.

algebra, 13, 16, 35, 84–85, 94, 98–99
altitude, 28–31
architecture, **8–9**, 36–39, **38**, **40**, 41, **42–43**, 44–47
assessment, 99
astronomy, 72–73, **73**, 75–77, **78–79**, **88**, 89, 97
axiom, 15

bank management, 105–109
baseline, 25
Bernoulli, Jakob, 16, **62**, 63
binary, 47, 50, 52–53, 56, 89
black holes, 72–73, **73**, 75–77, 97
Braess paradox, 109, 112–114
Brunelleschi, Filippo, **40**, 41

calories, 23–25, **23**
coefficient, 31
computer science, 47, **48**, 49–50, **51**, 52–53, **54**, 55–60

data, 89, 100–101, 104–15, 109
Descartes, Rene, 13, **14**, 15–16
differential equation, 94
domes, **40**, 41, 44–46
drag, 30–31

economics, **4**, 91, 105–109
Einstein, Albert, 19, 72–73, **74**, 75–76
encryption, 53, 56–58
Euclid, 10, 12–13, 15, 17, 19

Fibonacci series, 65–68
fluid, 30, 34, 82
fractals, 68–72, **69**

game theory, 87, 90–91
geometry, 10, 13, 17, 19, 35, 45, 68
golden ratio, 36–39, **37**, **38**, 68, **78–79**

interest, 63, 108–109, **108**
interpolation, 84
irregular, 17

joule, 22

kinetic, 22, **24**
Koch snowflake, 69–71, **69**
Konigsberg bridges, 116, **117**

Laplace, Pierre-Simon, 75, **96**, 97
Laplace transform, 94, 98–99
Lovelace, Ada, **54**, 55, 63

mass, 22, 31, 75–77, 92
minimal surfaces, **42–43**, 46–47
mountain climbing, 25–30

nursing, 34–35, **35**
nutrition, 23–25, **23**

optimize, 30

paradigm, 39
paradox, 11–12, 18, 109, 114
Parthenon, **8–9**, 36, **38**, 39
perpendicular, 31
plane, 16, 71
potential, 25
pressure, 27–28, 46
principal, **108**, 109
prisoners' dilemma, 87, 90–91
probability, 16, 63, 85, 89, 97
proposition, 17
psychology, 99–101, 104–105

ratio, 34, 36–38, 68, **78–79**, 82, 84
roof snow load, 39, 44

Schwarzschild radius, 76–77
search engine, 58–60, **58**
shepherds, **80**, 81–85
skydiving, 30–31, **32–33**, 36
Sum of Least Squares method, 101, 104–105
surface area, 31, 36, 39, 44–46

transcendental number, 61, 63
transportation planning, 109, 112–114

velocity, 22, 36, 92–93
video games, 91–94, **95**, 98–99

Zeno, 11–12, 16, 19

About the Author

Erik Richardson is an award-winning teacher from Milwaukee, where he has taught and tutored math up to the college level over the last ten years. He has done graduate work in math, economics, and the philosophy of math, and he uses all three in his work as a business consultant with corporations and small businesses. He is a member of the Kappa Mu Epsilon math honor society, and some of his work applying math to different kinds of problems has shown up at conferences, in magazines, and even in a few pieces of published poetry. As the director of Every Einstein (everyeinstein.org), he works actively to get math and science resources into the hands of teachers and students all over the country.